132. 50

Transportation Network Analysis

Transportation Network Analysis

Michael G. H. Bell
Transport Operations Research Group, University of Newcastle,
Newcastle upon Tyne, UK.

Yasunori Iida
Department of Transportation Engineering, University of Kyoto, Japan

JOHN WILEY & SONS
Chichester • New York • Weinheim • Brisbane • Singapore • Toronto

Copyright © 1997 by John Wiley & Sons Ltd,
Baffins Lane, Chichester,
West Sussex PO19 1UD, England

National   01243 779777
International (+44) 1243 779777
e-mail (for orders and customer service enquiries): cs-books© wiley.co.uk
Visit our Home Page on http//www.wiley.co.uk
or http//www.wiley.com

Other Wiley Editorial Offices

John Wiley & Sons, Inc., 605 Third Avenue,
New York, NY 10158-0012, USA

VCH Verlagsgesellschaft mbH, Pappelallee 3,
D-69469 Weinheim, Germany

Jacaranda Wiley Ltd, 33 Park Road, Milton,
Queensland 4064, Australia

John Wiley & Sons (SEA) Pte Ltd, 2 Clementi Loop #02-01,
Jin Xing Distripark, Singapore 129809

John Wiley & Sons (Canada) Ltd, 22 Worcester Road,
Rexdale, Ontario M9W 1L1, Canada

British Library Cataloguing in Publication Data

A catalogue record for this book is available from the British Library

ISBN 0 471 96493 X

Typeset in 10/12pt Times from the authors' disks by Laser Words, Madras, India
Printed and bound in Great Britain by Bookcraft (Bath) Ltd
This book is printed on acid-free paper responsibly manufactured from sustainable forestation,
for which at least two trees are planted for each one used for paper production.

Preface

It is of course trite to say that transportation networks are the arteries of the economy and that without them civilisation as we know it could not exist. Given the size and importance of the transport industry, the need for appropriate analytical techniques will be clear to all. Although there are many texts covering aspects of transportation network analysis, recent developments in the field mean that significant sections of this text are covering territory not covered by other books. In particular, the chapters on stochastic user equilibrium assignment, trip table estimation and network reliability present material not covered elsewhere.

The primary motivation for the book is to present a unified theoretical framework for transportation network analysis. The framework is an equilibrium one, because of the power and simplicity of the resulting methods. One of the characteristics of a transportation network is its size and complexity. A satisfactory network representation may require a large number of components (links and nodes) with a wide range of characteristics (length, capacity, type of control, etc.). Mathematical programming methods look attractive in this context, because they offer general methods for solving large problems and facilitate sensitivity analyses. A central portion of the book is therefore concerned with the formulation (and, to lesser extent, the solution) of equivalent optimisation problems and with sensitivity analysis.

An area of weakness for the equilibrium approach concerns congestion. At equilibrium there is a timeless balance between supply and demand, whereas congested conditions are characterised by transient overloads at particular points in the network. In order not to sacrifice the advantages of equilibrium methods, this book puts forward a compromise approach, whereby consecutive equilibrium time slices are linked by queues. The equilibrium queues in one time slice reduce the effective capacities of the corresponding links in the next time slice. In the next time slice, new equilibrium queues are only formed where the demand exceeds the effective capacity. The approach is therefore consistent with first-in, first-out (FIFO) behaviour at the link exits, which is where one would in any case expect to observe FIFO behaviour because of the possibility for overtaking before the link exit, at least in road networks. There is an inherent conflict in the assumptions, however, in that within each time slice all trips are assumed to be complete whereas the equilibrium queues constitute incomplete trips. The effect of these incomplete trips in subsequent time slices elsewhere in the network is not allowed for.

This book has benefited enormously from access to papers either in draft or at the refereeing stage. In particular, the introductory chapter benefited from ideas presented in a paper by Martin Mogridge, which at the time of writing had not been published. The compromise approach taken to time dependency described above owes much to as yet unpublished work produced by Ning Zhang, Hing Po Lo and William Lam.

1
Introduction

1.1 MOTIVATION AND APPROACH

There is a growing appreciation of the importance in developed societies of networks (or *lifelines*) of various kinds, such as water supply, energy supply, sewage disposal, communication and of course transportation (see, for example, Du and Nicholson, 1993). The vital importance of the transportation network is perhaps best appreciated in situations where it is severely disrupted, for example by an earthquake. A study of the effects of earthquakes on lifelines in the Wellington urban area of New Zealand considered their interdependence, and found that the transportation network is the most important. The restoration of other lifelines is dependent on people and equipment being able to move to the sites where damage has occurred, and the degradation of the transportation system inhibits this. Recent experiences in the Great Kobe Earthquake of 17 January 1995 confirm the importance of the transportation network in the rescue and restoration work.

Mobility and accessibility are major determinants of lifestyle and prosperity. At a macroscopic level, regions grow or decline as a result of changes in accessibility and forms of mobility. At a microscopic level, mobility and accessibility are major factors in business and residential location decisions. Increasing car ownership in the developed world has enabled people to move out of city centres into the suburbs (the so-called 'Flucht ins Grüne'). On the one hand, the greater dispersal and lower density of residences in suburban areas has fostered a growing car dependency. On the other hand, the increasing level of traffic congestion in urban areas is reducing the accessibility of many city centres. There is now a phenomenon in many countries of businesses relocating to more peripheral locations where there is more space and better access by road, leaving behind inner city ghettos and urban decay.

The trend to greater dispersal of both residences and businesses comes at a time when there is growing concern about the contribution transport is making to global pollution, and in particular to global warming. Many have argued that the most effective way to reduce the energy consumed, and the emissions produced, by the transport sector is to return to a less dispersed pattern of homes and jobs in order to reduce car dependency. This would require changes in lifestyle among at least some sections of the community. With the decline of 'smoke stack' industries (like coal, steel and shipbuilding), and the growth of cleaner forms of employment, the traditional functional separation into industrial areas and residential areas is becoming less desirable. A return to mixed use buildings, common in the pre-industrial era where the artisan lived above his workshop, is becoming increasingly feasible. If sections of the community can be persuaded to adopt more urban lifestyles, involving both living and working in city centres, then a step toward greater environmental sustainability would be achieved.

A rigorous consideration of the kinds of land use and transport issues just raised requires the appropriate tools to analyse transportation networks and travel behaviour within them. It is these tools that form the focus of attention of this book. In contrast to the more eclectic approaches taken by other texts in the field (for example, Ortuzar and Willumsen, 1990, or Sheffi, 1985), this book gives greater weight to coherence than to comprehensiveness. The techniques presented here are based on the notion of an *equilibrium* between the demand for, and the supply of, transportation. While a transportation system may never actually be in a state of equilibrium, it is assumed that the system is at least near equilibrium, tending toward equilibrium, and only prevented from attaining equilibrium by changes in external factors (referred to by economists as *exogenous variables*).

At equilibrium there is no incentive for any trip-maker to behave other than as he actually does. The demand for transportation is equal to its supply *at the prevailing cost to the user*. To use the jargon of economists, there is *market clearing*. The great attraction of equilibrium is that it is essentially timeless, so it is not necessary to look in detail at behavioural mechanisms, such as *how* or indeed *when* decisions are taken. In order to determine network flows, costs and other aspects of interest, it suffices under equilibrium just to look for those flows and costs that balance demand with supply. A further attraction of equilibrium is that, in general, the flows and costs that balance demand with supply may be formulated as the solution to a convex programming problem. This offers efficient algorithms for large networks.

In practice, as already noted, equilibrium does not apply exactly. For example, during a peak period more trips set off than can at that moment be accommodated by the network. There is a temporary excess of demand over supply, which is stored as one or more *queues* (of vehicles or passengers) to be discharged later. An essentially timeless representation of conditions is less useful. Queuing formulae based on equilibrium suggest that, as the demand for a facility (perhaps a road or a bus service) approaches its capacity, the expected queue and average delay tend to infinity. While this may be accurate over a very long period, demand exceeding capacity for a short period causes a finite queue (the shorter the period the shorter the queue). Because of the practical importance of congestion in transportation networks, some form of time dependency must be allowed for in the analysis.

A compromise approach is proposed in this book which retains the advantages of equilibrium theory while allowing for temporary overloading. Time is divided into slices. Equilibrium is assumed to prevail within each time slice, but queues are carried over from one time slice to the next. Within each time slice, each facility is confronted with the queue carried over from the preceding time slice plus the demand arising in the current time slice. Those flows (of vehicles, passengers or goods) which cannot be processed within the current time slice constitute the queue which is carried over to the next time slice, etc. This permits the formulation of convex programming problems while at the same time allowing the representation of the build up and decline of queues over, say, a peak period.

1.2 NETWORK REPRESENTATION

A network is referred to as a *pure network* if only its topology and connectivity are considered. If a network is characterised by its topology and flow properties (such as origin–destination demands, capacity constraints, path choice and link cost functions) it is then referred to as a *flow network* (see Du and Nicholson, 1993). A *transportation network* is a flow network representing the movement of people, vehicle or goods. Any transportation network can be represented as a *graph* in the mathematical sense, consisting of a set of *links* and a set of *nodes*. The links represent the movements between the nodes, which in turn represent points in space (and possibly also in time). The link may also refer to a specific *mode* of transport (for example, a movement by car, bus, train, bicycle or on foot), in which case a *path* in the transportation network specifies both the route and the mode(s) of transport. Subsequent chapters are concerned with network topology (*network design*), travel behaviour in networks (*assignment*), *network reliability* and flow monitoring (*path flow estimation*).

Travel behaviour within the network is governed by *costs*. Of particular interest is the *path cost*. At equilibrium, trip-makers choose the paths that they perceive at the time to be least cost. Where the network is appropriately specified, this may represent both route and mode choice. The *trip cost* is then equal to the cost of the path(s) chosen. The level of demand is determined by the trip costs. Underlying both path and trip costs are the *link costs*, as a path cost is the sum of the costs of the links constituting the path, and the trip cost is the cost of the path with the minimum perceived cost. Link cost is generally regarded to increase with link flow, although there is some debate about this (see Chapter 4). The relationship between link cost and link flow is called the *link cost function*.

1.3 MODES OF TRAVEL

The demand for transport is frequently characterised as a *derived demand*, arising out of economic, social or personal activities. Transportation requires not only a network in the sense of a physical infrastructure but also vehicles to carry the people, goods, or both. The exception is, of course, pedestrians who do not require vehicles. Two modes of transport may be identified. The first is *individual transport* (often referred to as private transport), consisting either of vehicles driven by the trip-makers themselves or of pedestrians. The second is *community transport* (frequently referred to as public transport), where the vehicles are driven by professionals. Within both the individual and community modes, a number of sub-modes are frequently identified. For example, motorised and non-motorised forms of transport in the case of the individual mode, or rail and bus in the case of the community mode.

Each mode (or sub-mode) of transport uses a network of some kind. Individual transport generally uses the road network, although in some instances there are cycle tracks and footpaths. Community transport may use the road or a rail network. Where mode choice is an aspect of the analysis, it may be convenient to define a transportation network that represents both individual and community transport. Where modes are combined in one network, it is important to specify the interchanges properly and associate with these the correct travel costs.

1.4 NETWORK EQUILIBRIUM

The concept of equilibrium has been touched on earlier. Under equilibrium, demand is equal to supply at the prevailing costs to the users. While the requirements for the

1.6 DEMAND CURVES

Conventionally it is assumed that demand curves slope downward, corresponding to a negative cost elasticity of demand. While in most cases, a fall in the price of a commodity is expected to increase (or at least not to reduce) the demand for it, a rather rare situation can arise with *inferior commodities* where the income effect of a price reduction dominates the substitution effect of a realignment of relative prices, leading to a demand curve that slopes upward. Such inferior commodities are referred to by economists as *Giffen goods*, after the Victorian economist who first identified the phenomenon. The oft quoted example of a Giffen good is that of potatoes during the Irish potato famine in 1845, when shortages pushed up the price making many families who were dependent on potatoes too poor to afford meat, thereby causing them to consume more, rather than less, potatoes (see, for example, Samuelson, 1970).

The recent RAC study of car dependency (RAC,1995) has produced cross-sectional evidence that the proportion of income devoted to travel by car rises with household income up to a certain level, after which point the proportion remains static. By contrast, the proportion of income devoted to travel by communal transport falls consistently with rising household income. It seems natural to conclude that the car is a superior mode for the lower income groups, while communal transport is an inferior mode for all income groups. Although it is conceivable that some households relying on community transport may use the saving engendered by a fall in fares to buy a car, leading to a reduced demand for community transport, there is no evidence that community transport is in fact a Giffen good.

While it seems incontrovertible that demand curves slope downward for all modes of transport (implying negative price elasticities), the total number of trips (as opposed to their length or timing) may prove to be rather inelastic to price changes (at least within normal ranges). For example, home-based trips are generated by *activities* occurring outside the home, so a reduction in the number of home-based trips would imply a reduction in the number of such activities or perhaps their relocation to within the home. While advances in telecommunications may allow some basic activities, like working, education or shopping, to occur at home, there is no evidence that this will occur on a significant scale in the near future. There is a fundamental human need to gather periodically. Over a long period of time, in fact since 1929, it has been found that the number of trips made per person has remained reasonably constant at around one thousand per year (see Köhl, 1994).

The amount of travel (measured in, say, passenger-kilometres or tonne-kilometres) is, however, likely to be sensitive to cost. In the case of work trips or freight movements, it is of course possible for *supply chains* to change in response to changes in transport costs, leading to a substitution of transport for other inputs in the process of production. The amount of transport input can be influenced by decisions relating to local sourcing, just-in-time deliveries, economies of scale, etc. Similar arguments apply to non-work trips.

1.7 LINK COST FUNCTIONS

Physics has two theories of light, namely wave theory and quantum theory, both of which yield useful and verifiable predictions, but are not entirely consistent with each other. So too there are two theories of traffic, hydrodynamic theory and queuing theory, both having their uses but differing in their predictions. Hydrodynamic theory treats traffic as a compressible fluid. It predicts shock waves and suggests two speeds for any flow other than maximum flow and zero flow. Queuing theory quantises traffic and concentrates on the arrival and departure processes at particular points, like stop lines or bus stops. The theories differ in their predictions of the shape of the link cost function. According to hydrodynamic theory, the link cost function bends backward after maximum flow is attained, suggesting two costs for any flow other than the maximum flow and zero flow. In contrast, queuing theory predicts monotonically increasing link costs.

The fundamental diagram (see Fig. 1.1) underlies the hydrodynamic theory of traffic as originally formulated by Lighthill and Whitham (1955). Traffic is treated as a compressible fluid with zero mass (second order fluid models relax the zero mass assumption). Note that dimensional consistency requires that flow (measured, say, in vehicles per hour) is equal to speed (measured, say, in kilometres per hour) multiplied by density (measured, say, in vehicles per kilometre). Zero flow is encountered at zero density and again at maximum or queue density. In between zero and queue density, flow and speed are non-zero. It follows from common sense (and indeed the Intermediate Value Theorem) that between zero and queue density there is a density and a corresponding non-zero speed at which flow is at a maximum. Provided that flow is a continuous function of density (in other words, that there are no densities for which flow is undetermined), and provided that there are no local maxima, then for any flow other than the maximum, there are two possible speeds, a lower speed corresponding to the higher density and a higher speed corresponding to the lower density.

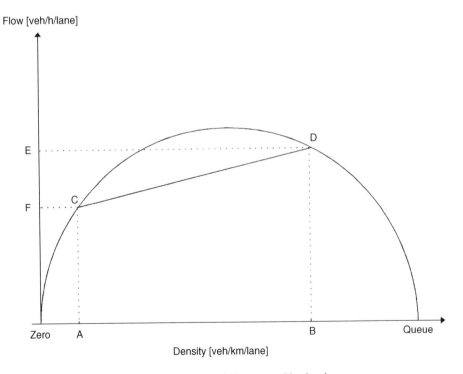

Figure 1.1 *Fundamental diagram with shock wave*

Speed depends only on density in accordance with the fundamental diagram. When density changes, speed changes simultaneously. This implies infinite acceleration or deceleration, hence the assumption of zero mass. Consider a link with a given flow. Flow conservation at any point along the link requires that where traffic at two different densities encounter each other, the boundary between the two (referred to as the *shock wave*) moves at a speed given by the slope of the chord that connects the two points on the fundamental diagram defined by the two densities. Suppose traffic at density A is followed by traffic at density B. Then, as Fig. 1.1 shows, a shock wave is generated which travels with the speed given by the slope of the chord CD. The direction of travel of the shock wave depends on the densities A and B, and can be either positive or negative.

Shock waves are concerned with the dynamical properties of traffic on links. As this book concentrates on network equilibrium, the dynamical properties of traffic are not of direct interest. Of interest is the relationship between speed and flow. As flow at equilibrium is equal to speed times density, the speed–flow curve may be inferred directly from the fundamental diagram. Note that in general there are two speeds for every flow, the higher speed corresponding to the lower density and the lower speed corresponding to the higher density. The generalised cost of an uncontrolled link will in general be related to flow because speed is related to flow and travel time is inversely related to speed. This leads to the phenomenon of the backward bending cost curve (see Fig. 1.2).

There is some controversy over whether for a given flow there are in fact two stable equilibria. The higher cost corresponding to congested conditions is likely to be transitory as congestion builds (or declines). Measurements of speeds and flows made on expressways (see, for example, Leutzbach, 1988) exhibit a stable relationship at the lower density

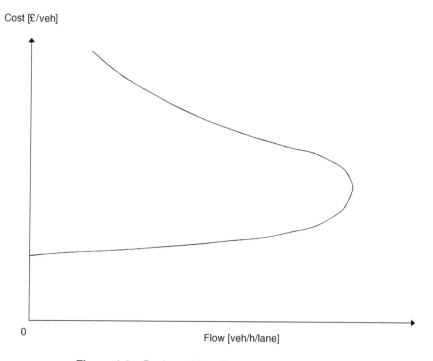

Figure 1.2 *Backward bending cost curve for a link*

(the lower branch of the cost–flow curve) but chaotic behaviour at the higher density (the upper branch of the cost–flow curve).

Hydrodynamic theory is most applicable to long, uncontrolled links, like expressways. For links where the dominant effect of flow on cost occurs at junctions, bus or tram stops, etc., queuing theory is likely to be more applicable. Usually (and certainly in this book) the network is specified so that each link has at most one queuing process located at the exit. Of particular interest are links where the service has a cyclical character, for example at traffic signals, where vehicles are only serviced during green, or at bus or tram stops, where passengers are serviced only when a bus or tram arrives.

Link travel time may be decomposed into delay and undelayed time, where delay increases with flow. Delay formulae play a central role in the determination of optimal traffic signal settings. The first well-publicised formula, proposed by Webster (1958), divided delay into a uniform component and a random component. Webster's formula relates to steady state networks, so delay is only defined while demand does not exceed capacity, where in this case capacity is determined by the green time and the saturation flow (the rate at which a queue discharges). For links where the service is random, like roads leading into give way junctions, the Pollaczek–Khintchine formula may be used (see Chapter 4 for further details).

In reality, of course, the link entry flow may temporarily exceed the link exit capacity, causing a queue to form. Time-dependent formulae (see, for example, Kimber and Hollis, 1979) relate to *time slices* and allow for temporary overloading. A property of links with queuing processes is that delay increases monotonically with flow, leading to monotonically increasing link cost functions that do not bend backwards (see Fig. 1.3).

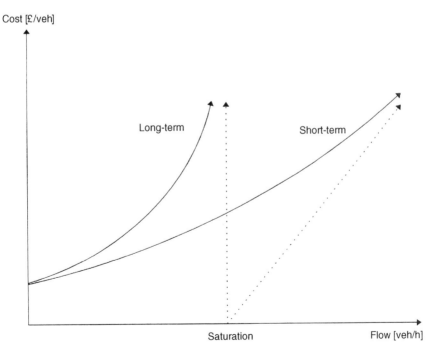

Figure 1.3 *Monotonic cost curves for a link with a queuing process*

1.8 SUPPLY CURVES

In the case of individual mode trips, the effect of increasing the trip rate between an origin and a destination will be to *increase* the trip cost, provided the capacity of the network has not been reached. The problem of backward bending cost curves is less likely to be encountered when sufficient alternative paths are available, because traffic can divert if the congestion on the current paths makes alternative paths more attractive. When two paths connect a given origin to a given destination, the trip cost resembles the curve depicted in Fig. 1.4 if trip-makers are user optimising and know the path costs accurately. At zero flow, path 1 is cheaper. As flow is increased, the cost of path 1 increases. When the cost of path 2 is just exceeded, sufficient traffic will divert to path 2 to ensure that the costs of the two paths remain equal (note that the length of AB is equal to the length of CD). The deterministic user equilibrium trip cost curve is therefore rather flatter than either path cost curve and also flatter than the cost curve obtained by assigning a fixed proportion of traffic to either path.

Another way to view the deterministic user equilibrium between the two paths is to superimpose the two cost curves as shown in Fig. 1.5. The distance between the left and right vertical axes corresponds to the origin-to-destination trip rate. To the left of the intersection of the two curves, path 1 costs less than path 2 and there is an incentive to swap to path 1. To the right of the intersection, path 2 costs less than path 1 and there is an incentive to swap to path 2. The equilibrium is therefore stable, provided both path cost functions are monotically increasing in path flow. If the two curves are pulled apart horizontally, corresponding to an increase in the trip rate, the deterministic user

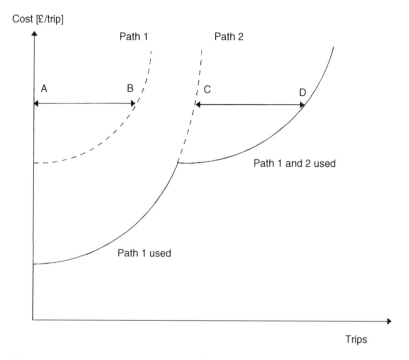

Figure 1.4 *Trip cost under deterministic user equilibrium with two paths*

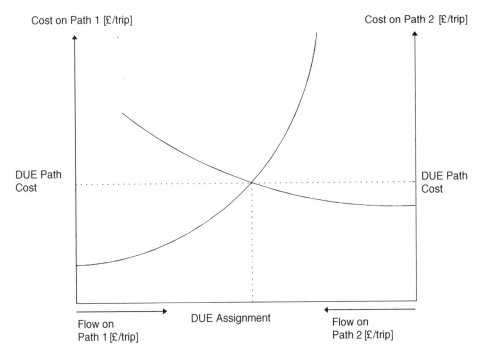

Figure 1.5 *Deterministic user equilibrium for two paths*

equilibrium trip cost increases. The relationship between trip rate and trip cost constitutes the supply curve.

In the case of communal transport, the operator(s) come between the users and the network, so the costs faced by the operators do not necessarily reflect accurately the prices charged to users. Indeed, operators often seek to maximise revenue, charging users more where 'the market will bear it', namely where demand is relatively price inelastic. The relationship between flow and cost experienced by the user is therefore less direct than for individual transport. In the following, it is assumed that the fare per trip tends to follow the cost per trip to the operator. To the fare per trip must be added the value of the time spent waiting and travelling, plus the value of any other significant factors, in order to arrive at the cost per trip. The resulting cost per trip when plotted against the number of trips constitutes the supply curve for communal transport.

Most forms of communal transport operate fixed routes. The user, who wishes to get from one point on the network (call it the origin) to another (call it the destination), may have a choice of *line*, or combination of lines, which may or may not take different physical paths. The effect of passenger flow on travel time is through waiting and boarding times. An increase in passenger flow will *ceteris paribus* result in an increase in waiting time, because the probability that any passenger taken at random will not be able to get into the next appropriate vehicle will increase. However, if the operator(s) increases service frequencies in response to an increase in passenger flow, waiting time may well fall. There is a point beyond which service frequencies cannot be further increased due to insufficient network capacity, leading to an increasing relationship between passenger flow and waiting time.

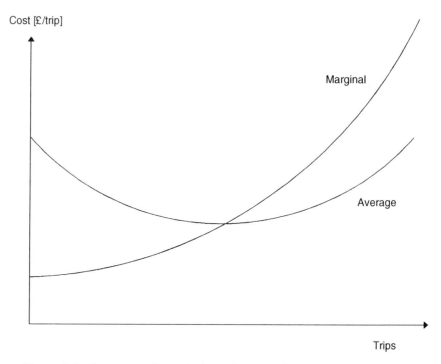

Cost [£/trip]

Marginal

Average

Trips

Figure 1.6 *Average and marginal supply curves for communal transport*

The generalised cost per trip for communal transport is therefore initially likely to fall with increasing flow, because the fixed component of cost is shared over more trips and service frequencies can be increased, thereby reducing waiting times. Beyond a certain point, however, the capacity of the network is approached and the trip cost begins to rise. Since the trip cost experienced by the user is an average cost, the marginal cost is less than the average cost for lower levels of flow and greater than the average cost for higher levels of flow, as shown in Fig. 1.6. Different modes of communal transport will of course have different supply curves, but the pattern would be expected to be the same.

1.9 INTERMODAL EQUILIBRIUM

The shape of the supply curve for communal transport has some interesting consequences. The first, referred to as the Downs–Thompson paradox, after Downs (1962) and Thompson (1977), is that investment in additional road capacity for individual transport may increase trip cost. Under a deterministic user equilibrium, the trip cost between any given origin and destination by either individual or communal transport should be equal if both modes are used. An increase in road capacity attracts some traffic away from communal transport to individual transport. This may increase the average cost of travel by communal transport, because reduced demand means reduced frequencies of service and increased waiting times. The new equilibrium will be at a higher trip cost. The paradox is illustrated in Fig. 1.7.

When the trip rate is fixed (is inelastic), the supply curves for individual and communal transport can be plotted against each other. The point where the curves intersect defines

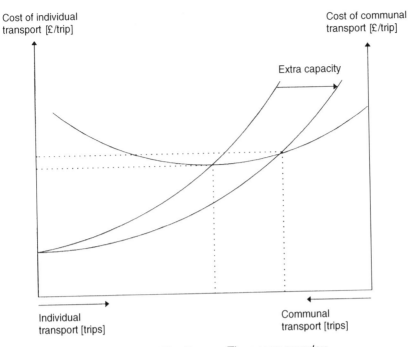

Cost of individual
transport [£/trip]

Cost of communal
transport [£/trip]

Extra capacity

Individual
transport [trips]

Communal
transport [trips]

Figure 1.7   *The Downs–Thompson paradox*

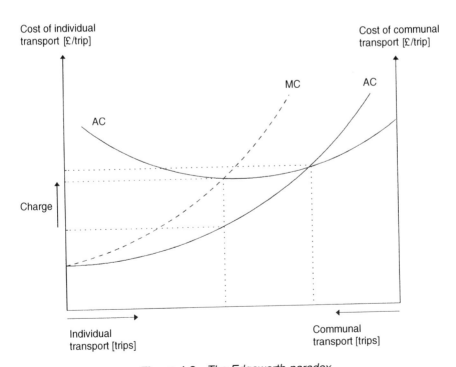

Cost of individual
transport [£/trip]

Cost of communal
transport [£/trip]

MC   AC

AC

Charge

Individual
transport [trips]

Communal
transport [trips]

Figure 1.8   *The Edgeworth paradox*

the deterministic user equilibrium. By considering the incentives generated by deviating from the equilibrium it can be inferred that the equilibrium is stable. When the capacity of the road network is increased, the supply curve for individual transport shifts to the right. The new point of intersection in Fig. 1.7 implies a shift away from communal transport toward individual transport and a higher equilibrium trip cost.

Another curious effect, known as the Edgeworth paradox and pointed out in this context by Mogridge (1995), is that the introduction of a tax on individual transport can reduce the equilibrium trip cost. As illustrated in Fig. 1.8, a tax (in this case equal to the difference between the average and marginal cost) leads to a new equilibrium with a shift away from individual transport to communal transport and a lower trip cost. The introduction of a congestion toll could therefore *reduce* equilibrium trip costs.

1.10 USER BENEFIT AND SURPLUS

When the trip rate is inelastic, as in the preceding section, the user equilibrium price is determined by the supply curve(s) alone. However, demand is usually elastic (price sensitive), leading to the introduction of the trip demand curve. The intersection of the demand and supply curves determines the user equilibrium costs and flows, as illustrated in Fig. 1.9. The *user benefit* is the amount that users would be willing to pay for a given level of trip-making, given by the area OAEB. Note that unless there is a cost at which demand is zero or unless demand tends to zero sufficiently rapidly as cost tends to

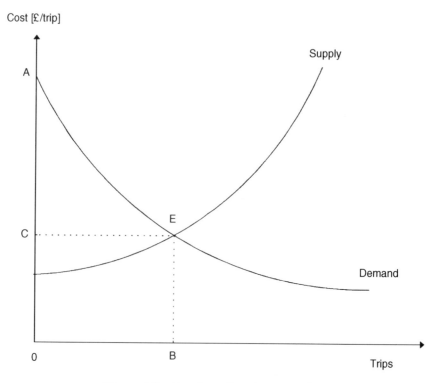

Figure 1.9 *User benefit and user surplus*

2
Transportation Networks

2.1 INTRODUCTION

In order to predict how the demand for mobility will be manifested in space and time, it is necessary to represent the transportation infrastructure in some formal, simple but sufficiently detailed way. The approach adopted almost universally is to represent the infrastructure by a set of links and a set of nodes. The relationship between the links and the nodes, referred to as the *network topology*, can be specified by a *link-node incidence matrix*. This is one of the few areas of transportation modelling where there is a near-unanimity of approach. While attempts to develop continuum models are periodically made (see Yang *et al.*, 1994, for a recent attempt), these have yet to replace network-based models, principally because of their inability to represent the properties of a transportation infrastructure in sufficient detail. This chapter sets out the principles of the network-based representation of transportation systems.

2.2 NETWORK TERMINOLOGY

A transport network may be formally represented as a set of *links* and a set of *nodes*. A link connects two nodes and a node connects two or more links. Links may be either *directed*, in which case they specify the direction of movement, or *undirected*. Two links are said to be *parallel* if they connect the same pair of nodes in the same direction. A *loop* is a link with the same node at either end. Some of the algorithms described later require the exclusion of parallel links and loops. All links referred to in later chapters are directed.

Links may have various characteristics. In the context of transportation network analysis, the following are some of the characteristics of interest:

- *link length* (in metres or perhaps in average vehicles);
- *link cost* (sometimes travel time but more generally a linear combination of time and distance); and
- *link capacity* (maximum flow).

A link may be regarded as a conduit for flow whose units of measurement will depend on the application (for example, vehicles per hour or passengers per hour). Flow consists of one or more *commodities*. While the commodities may refer to different kinds of goods and services, in this context they often refer to other aspects of a unit of flow, like its *origin* or its *destination* (sometimes referred to as the *source* and *sink* respectively).

If there is only one origin and one destination, and no other relevant classification, flow is said to be *single commodity*. In addition to commodities, *user classes* are sometimes identified to distinguish between units of flow with different forms of *travel behaviour*.

A *movement* in a transportation network corresponds to a flow with a distinct origin and destination. Origins and destinations may correspond to specific buildings, like a house or an office, or to zones, depending on the level of aggregation. From the perspective of a transportation network, an origin or destination is represented by a kind of node, referred to as a *centroid*. Each centroid is connected to one or more *internal nodes* by a kind of link referred to as a *centroid connector* (or just a *connector*). While links tend to correspond to identifiable pieces of transport infrastructure (like a section of road or railway), centroid connectors are artefacts, especially when the respective centroid corresponds to a zone with in reality multiple entrances and exits.

Figure 2.1 provides an example of a transportation network with one origin centroid, two destination centroids, five links, four internal nodes, three connectors and 5 paths. Note the conventions used to represent the centroids, connectors, internal nodes and links.

There is a parallel branch of mathematics known as graph theory which uses a terminology which differs to that used by transportation planners or traffic engineers. A transport network could be referred to as a *valued graph*, or alternatively a *network* (as here), or a *net*. Directed links are referred to as *arcs* while undirected links (not encountered in this book) are referred to as *edges*. Much of the literature on transport networks emanating from operations researchers and mathematicians tends to use graph theory terminology. The terminology adopted here is that generally used by transportation planners and traffic engineers.

Other useful terms with mostly intuitive interpretations (see Fig. 2.2) are: a *path*, which is a sequence of distinct nodes connected in one direction by links; a *cycle*, which is a path connected to itself at the ends; a *tree*, which is a network where every node is visited once and only once; and a *cutset*, which is a minimal collection of links whose removal from the network would cut the network in two with no links between the two resulting sub-networks.

Network concepts are summarised in Table 2.1.

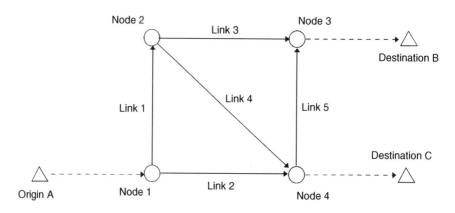

Figure 2.1   *An example network*

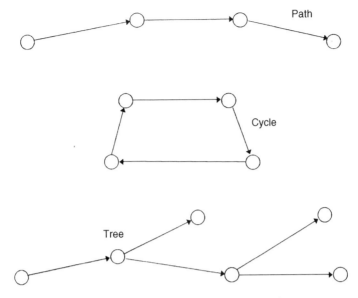

Figure 2.2   *Example of a path, a cycle and a tree*

Table 2.1   *Network concepts and their definitions*

Concept	Definition
Node	Junction of two or more links. It is either an *internal node* (neither source nor sink) or a *centroid* (either source or sink)
Link	Conduit for flow between two nodes
Connector	Link between a centroid and an internal node
Movement	Flow with a specified origin and destination
Path	A sequence of nodes connected by links in one direction so that a movement is feasible from the first node to the last node in the sequence. Often the first and last node in the sequence are centroids
Cycle	A path with the same node at either end
Tree	A network where each node can be visited once and only once
Cutset	A set of links whose removal would lead to two sub-networks with no connections
Commodity	A kind of flow distinguished by origin, destination or some other feature
User class	A kind of flow distinguished by its behaviour, for example its sensitivity to cost

2.3   TRANSPORTATION NETWORK TYPES

Some network topologies are commonly encountered in transportation systems. One such is the *linear network*, characterising perhaps an expressway, an arterial road or a railway line. There may be many origins and destinations but no choice of path. Another is the *grid network*, representing perhaps an urban area consisting of blocks as is common in the USA and Japan. There may be many origins and destinations as well as many alternative routes.

and j, and number paths distinctly and consecutively, producing the path flow vector

$$\mathbf{h}^{\mathrm{T}} = [h_1, h_2, \ldots, h_P]$$

where P is now the total number of paths. When considering time-dependent or multi-commodity flows, additional subscripts may be introduced. For example, when link flows are distinguished by origin and destination as follows:

$$v_{ijk} = \text{the flow on link } k \text{ between origin } i \text{ and destination } j$$

When the number of origins and/or destinations exceeds one, the flow on link k is multi-commodity, in the sense described earlier.

The origin-to-destination flows, referred to collectively as the *trip table*, may be conveniently represented in vector form as follows:

$$\mathbf{t}^{\mathrm{T}} = [t_{11}, t_{12}, t_{13}, \ldots, t_{1J}, \ldots, t_{IJ}]$$

where I is the number of origins and J is the number of destinations.

The following notation is adopted for cost variables,

$c_k = $ the unit cost of travel on link k;

$g_{pij} = $ the unit cost of travel on path p between origin i and destination j; and

$z_{ij} = $ the unit cost of a trip from origin i to destination j.

As with path flows, the origin and destination subscripts are usually omitted from path cost variables and the paths numbered distinctly and consecutively. In vector and matrix notation

$$\mathbf{c}^{\mathrm{T}} = [c_1, c_2, \ldots, c_K]$$
$$\mathbf{g}^{\mathrm{T}} = [g_1, g_2, \ldots, g_P]$$
$$\mathbf{z}^{\mathrm{T}} = [z_{11}, z_{12}, z_{13}, \ldots, z_{1J}, \ldots, z_{IJ}]$$

A glossary of notation is provided at the end of Chapter 3.

2.5 INCIDENCE MATRICES

An *incidence matrix* is a table of binary or ternary variables stating the presence or absence of a relationship between network elements and other variables. The incidence matrix specifies the network topology. A useful matrix for network processing is the *node–link incidence matrix* with elements

$$e_{nk} = \begin{cases} \;\;\,1 & \text{if node } n \text{ lies at the exit of link } k \\ -1 & \text{if node } n \text{ lies at the entrance of link } k \\ \;\;\,0 & \text{otherwise} \end{cases}$$

The node-link incidence matrix has the form

$$\mathbf{E} = \begin{bmatrix} e_{11} & e_{12} & \cdots & e_{1k} \\ e_{21} & e_{22} & \cdots & e_{2k} \\ & & \cdots & \\ e_{N1} & e_{N2} & \cdots & e_{NK} \end{bmatrix}$$

Two incidence matrices frequently encountered in this book are the *link–path incidence matrix* and the *origin–destination–path incidence matrix*.

The link–path incidence matrix has elements

$a_{kpij} = 1$ if path p from origin i to destination j uses link k, and 0 otherwise

Dropping the subscripts i and j, and numbering the paths distinctly and consecutively, yields

$a_{kp} = 1$ if path p uses link k, and 0 otherwise

The link–path incidence matrix then has the form

$$\mathbf{A} = \begin{bmatrix} a_{11} & a_{12} & \cdots & a_{1P} \\ a_{21} & a_{22} & \cdots & a_{2P} \\ & & \cdots & \\ a_{K1} & a_{K2} & \cdots & a_{KP} \end{bmatrix}$$

The origin–destination–path incidence matrix has elements

$b_{ijp} = 1$ if path p connects origin i with destination j, and 0 otherwise.

and has the following form

$$\mathbf{B} = \begin{bmatrix} b_{111} & b_{112} & \cdots & , b_{11P} \\ b_{121} & b_{122} & \cdots & b_{12P} \\ & & \cdots & \\ b_{IJ1} & b_{IJ2} & \cdots & b_{IJP} \end{bmatrix}$$

Taking Fig. 2.1 as an example, the node–link incidence matrix is

$$\mathbf{E} = \begin{bmatrix} -1 & -1 & 0 & 0 & 0 \\ 1 & 0 & -1 & -1 & 0 \\ 0 & 0 & 1 & 0 & 1 \\ 0 & 1 & 0 & 1 & -1 \end{bmatrix}$$

Furthermore, the link–path incidence matrix is

$$\mathbf{A} = \begin{bmatrix} 1 & 1 & 0 & 1 & 0 \\ 0 & 0 & 1 & 0 & 1 \\ 1 & 0 & 0 & 0 & 0 \\ 0 & 1 & 0 & 1 & 0 \\ 0 & 1 & 1 & 0 & 0 \end{bmatrix}$$

when path 1 uses links 1 and 3, path 2 uses links 1, 4 and 5, path 3 uses links 2 and 5, path 4 uses links 1 and 4, and path 5 uses link 2. The corresponding origin–destination–path incidence matrix is

$$\mathbf{B} = \begin{bmatrix} 1 & 1 & 1 & 0 & 0 \\ 0 & 0 & 0 & 1 & 1 \end{bmatrix}$$

since paths 1, 2 and 3 connect centroid A to centroid B and paths 4 and 5 connect centroid A to centroid C.

2.6 *CONSERVATION RELATIONSHIPS*

As stated earlier, it is assumed that flow enters and leaves the network only by the centroids. At all internal nodes flow is conserved. This means that the flow into each internal node must equal the flow out of it, or in the case of multicommodity flows, the flow of *each commodity* into each internal node must equal the flow *of the same commodity* out. For each internal node n therefore there is a relationship of the form

$$\mathbf{e}_n^{\mathrm{T}}\mathbf{v} = 0$$

where $\mathbf{e}_n^{\mathrm{T}}$ is row n of the node-link incidence matrix defined earlier. This implies that for each node, one of the associated flows is linearly dependent on the other flows.

Returning to Fig 2.1, there is one internal node (node 2) for which

$$[1 \quad 0 \quad -1 \quad -1 \quad 0]\mathbf{v} = 0$$

(the rules of vector and matrix multiplication are given in Chapter 3). Representing the flows on connectors 1, 2 and 3 as v_6, v_7 and v_8 respectively produces four conservation constraints, one for each node, namely

$$
\begin{bmatrix}
-1 & -1 & 0 & 0 & 0 & 1 & 0 & 0 \\
1 & 0 & -1 & -1 & 0 & 0 & 0 & 0 \\
0 & 0 & 1 & 0 & 1 & 0 & -1 & 0 \\
0 & 1 & 0 & 1 & -1 & 0 & 0 & -1
\end{bmatrix}
\mathbf{v}' =
\begin{bmatrix}
0 \\
0 \\
0 \\
0
\end{bmatrix}
$$

where $\mathbf{v}'$ is the link flow vector extended to include the connector flows.

In the case of multicommodity flows, the above relationship applies for each commodity. For example, when flows are distinguished by their origins and destinations

$$\mathbf{e}_n^{\mathrm{T}}\mathbf{v}_{ij} = 0$$

where $\mathbf{v}_{ij}$ is a vector of link flows from origin i to destination j. This implies that for each commodity and each node, one of the flows in or out of the node is linearly dependent. In the case of node 2 in Fig. 2.1, this implies that

$$[1 \quad 0 \quad -1 \quad -1 \quad 0]\mathbf{v}_{ij} = 0 \qquad \text{for } i = A \quad \text{and} \quad j = B \text{ or } C$$

Conservation relationships have a number of other forms. One form frequently encountered in this book is the following:

$$\mathbf{v} = \mathbf{A}\mathbf{h}$$

which asserts that the flow on any link is equal to the sum of the flows on all of the paths using that link. In the case of Fig. 2.1,

$$
\mathbf{v} = \mathbf{A}\mathbf{h} =
\begin{bmatrix}
1 & 1 & 0 & 1 & 0 \\
0 & 0 & 1 & 0 & 1 \\
1 & 0 & 0 & 0 & 0 \\
0 & 1 & 0 & 1 & 0 \\
0 & 1 & 1 & 0 & 0
\end{bmatrix}
\mathbf{h} =
\begin{bmatrix}
h_1 + h_2 + h_4 \\
h_3 + h_5 \\
h_1 \\
h_2 + h_4 \\
h_2 + h_3
\end{bmatrix}
$$

Another conservation relationship encountered in this book is the following:

$$t = Bh$$

which, in the case of Fig. 2.1, has the form

$$t = Bh = \begin{bmatrix} 1 & 1 & 1 & 0 & 0 \\ 0 & 0 & 0 & 1 & 1 \end{bmatrix} h = \begin{bmatrix} h_1 + h_2 + h_3 \\ h_4 + h_5 \end{bmatrix}$$

This asserts that the flow between any origin and destination is equal to the sum of the flows on all the paths connecting that origin to that destination.

A further useful conservation relationship is

$$v = Pt$$

where P is a matrix of *link choice proportions* (or, looking at the relationship stochastically, *link choice probabilities*). The derivation of P is discussed later.

Conservation relationships may be regarded as the plumbing of a model. The connections are made in such a way as to not lose flows or, where required, not to mix commodities or user classes.

2.7 SHORTEST PATH ALGORITHMS

A crucial step in many network flow programming methods is the finding of shortest (or, more generally, least cost) paths between any pair of nodes or centroids. Two shortest path algorithms are described in this section, each with a conceptually distinctive approach. Both the algorithms presented here operate on a *connections matrix* and a *backnode matrix*. Let N be the number of nodes and Z the number of centroids. The connections matrix is a matrix with $N + Z$ rows and $N + Z$ columns whose elements are c_{mn}. It is initialised as follows: Where there is a link or centroid connector from node or centroid m to node or centroid n, c_{mn} is assigned its cost; where no such link or centroid connector exists, c_{mn} is assigned a large value.

During processing, connections are made and, if new or better than the preceding connection, entered in the connections matrix. The algorithms differ with respect to the way in which the connections are sequenced. After processing, the connections matrix gives the minimum cost of travelling from any node or centroid to any other node or centroid. When c_{mn} retains its preassigned large value, node or centroid n is not reachable from node or centroid m.

The backnode matrix has the same size as the connections matrix. After processing, element q_{mn} of this matrix gives the penultimate node on the path from node or centroid m to node or centroid n. The backnode matrix enables the optimal path to be traced. Suppose that $q_{mn} = k$, then the optimal path from m to n has as its penultimate node k, and the node preceding k is given by q_{mk}, etc.

A method that is simple to program is that due to Floyd (1962) and Warshall (1962). The Floyd–Warshall algorithm can be stated as follows, using a BASIC-like language.

Floyd—Warshall shortest path algorithm

Step 1   (initialisation of the connections and backnode matrices)
            for all nodes or centroids m and n

$q_{mn} \leftarrow m$
if a link joins m to n then $c_{mn} \leftarrow$ link cost
otherwise $c_{mn} \leftarrow \infty$

Step 2   (obtain least cost paths)
for all nodes k
    for all nodes or centroids m not equal to k
        for all nodes or centroids n not equal to k or m
           if $c_{mk} + c_{kn} < c_{mn}$ then $c_{mn} \leftarrow c_{mk} + c_{kn}$ and $q_{mn} \leftarrow q_{kn}$

As the iterations progress, paths are formed, and when parallel paths emerge, the lower of the two costs is retained. The algorithm works because all paths and sub-paths are generated.

To illustrate the algorithm, consider Fig. 2.1. Suppose $A \to 1 \to 2 \to 4 \to 3 \to B$ is the optimal path from A to B. At some stage during the first outer loop (when $k = 1$), the connection $A \to 2$ is made. This is followed in the second outer loop (when $k = 2$) by the optimal connection $A \to 4$ via nodes 1 and 2, and in the third outer loop (when $k = 3$) by the connection $4 \to B$. Finally, in the fourth outer loop (when $k = 4$) the optimal connection $A \to B$ is established. If the nodes were numbered differently, a different series of connections would be made but the end result would be the optimal connection $A \to B$.

The Floyd–Warshall algorithm is attractive for networks where all (or most) nodes are connected initially and where least cost paths are required between all nodes, not just between centroids. In transportation networks the ratio of links to nodes is typically around only 3, according to Van Vliet (1978). An alternative, more efficient approach to finding least cost paths is to build a least cost path tree, as is done by the algorithm due to Dijkstra (1959). Let V be the set of nodes visited so far. Dijkstra's algorithm may be stated as follows.

Dijkstra's shortest path algorithm

Step 1   (initialisation of the connections and backnode matrices)
for all nodes or centroids m and n
    $q_{mn} \leftarrow m$
    if a link joins m to n then $c_{mn} \leftarrow$ link cost
    otherwise $c_{mn} \leftarrow 0$

Step 2   (obtain least cost paths)
for all centroids m
    $k \leftarrow m$
    set V is empty
    set E is empty
    repeat
        add k to set V
        for all n connected to k
           put n in set E
           if $c_{mk} + c_{kn} < c_{mn}$ then $c_{mn} \leftarrow c_{mk} + c_{kn}$ and $q_{mn} \leftarrow q_{kn}$
        select k in E and not in V so that c_{mk} is minimum
    until V contains all nodes

Set V contains the set of visited nodes; note that every node is visited only once. Set E contains the nodes reached so far by branching, not all of which are eligible for selection. Each iteration of the innermost loop adds a branch to the tree and checks whether any node has thereby been reached with less cost. If so, the minimum cost and backnode matrices are updated.

The operation of the algorithm in relation to Fig. 2.1 is illustrated in Fig. 2.5 for hypothetical link costs. When the least cost path tree is two links deep, node 2 is selected for branching as it is the node reached so far (namely in set E) that is not in set V and is closest to A (the route of the tree). Node 2 is added to set V. From node 2, nodes 3 and 4 are reached and therefore added to set E. The node next selected for branching is node 4, as it is in set E, has least cumulative cost and is not in set V. Node 4 is added to the set V, which ensures that the path $A \rightarrow 1 \rightarrow 4$ is not developed further. This path would clearly be sub-optimal, as node 4 is reached at less cost via nodes 1 and 2.

The optimality of Dijkstra's algorithm is guaranteed by the selection mechanism for k, the node for branching. Each node k is selected only once. When node k is selected, it must lie on an optimal path; if not, k must be reachable with less cost, in which case it would have been selected earlier.

Either Dijkstra's algorithm or an efficient dynamic programming algorithm (not described here) will be the most efficient for transport networks (see Van Vliet, 1978). Dijkstra's algorithm, unlike Floyd's algorithm, works only with positive costs, although in the context of transportation network analysis this is not usually a limitation. No algorithm will of course work where negative cycles are present, because a negative cycle would imply that the trip cost could be reduced *ad infinitum*.

Consider again the network in Fig. 2.1 and assume that the connections matrix before processing by one of the algorithms is as given in Table 2.2. The entry 99999 indicates that there is no link from the given node/centroid to the given node/centroid. The corresponding backnode matrix before processing is as given in Table 2.3.

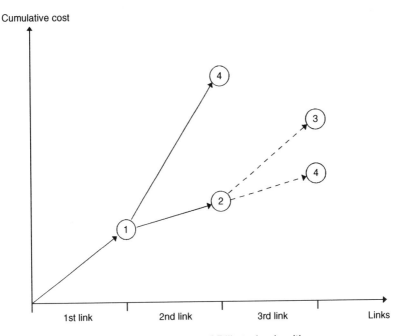

Figure 2.5 *Operation of Dijkstra's algorithm*

2.11 NETWORK CAPACITY

It seems reasonable to suppose that every link has a maximum flow that it can carry, referred to here as its *capacity*. In the case of roads, this is often taken to be the rate at which a queue of cars discharges over a *stop line*, which has been found to be about 2000 cars per hour per lane, with variations between countries. It is worth noting in passing that significantly higher flow rates have been recorded on expressways in the USA.

The capacity of a path is the minimum of the capacities of the constituent links. For a network with one origin and one destination and one user class, the capacity is equal to the minimum of the cutset capacities (according to the *max-flow min-cut theorem*). Ford and Fulkerson (1962) proposed a method for finding this capacity.

For more general networks, with multiple origins, destinations or user classes, there is a problem in defining what is meant by capacity. There is usually scope for increasing the flow of one commodity at the expense of another. In other words, alternative definitions of capacity are possible for multicommodity flow. Suppose that flows are assigned to the least cost routes and that links have finite capacities. One definition of capacity might in this case be the maximum feasible multiplier that can be applied to a trip table. Network capacity will change if either the trip table or the assignment rule is changed. In particular, the capacity will tend to decrease as the proportion of longer trips increases, since longer trips will tend to use more links.

Traffic control on urban expressways and the determination of land-use capacities for a given transportation network are practical examples of the application of network capacity concepts. In the former case, the problem is to maximise the total on-ramp inflows of an urban expressway network for a given set of on-ramp-to-off-ramp choice probabilities and subject to link capacities. In the latter case, the problem is to obtain the upper limit for the land-use development of each zone subject to the link capacities. One can distinguish two approaches, one where the destination choice probabilities are fixed and the other where they are related to network characteristics. The related concept of network reliability is discussed in more detail in Chapter 8.

For signal setting at junctions, capacity plays an important role (see Chapter 4). Allsop (1992) has defined the practical capacity as the largest common multiple of the arrival rates for which signal timings can be found that satisfy all the safety and capacity constraints. There is, however, the possibility to gain more capacity for some stream at the expense of the conflicting streams. Yang and Wong (1995) have extended the approach of Allsop (1992) to networks under equilibrium assignment. The largest common multiple of the trip table is sought while not exceeding the maximum degree of saturation on any signal controlled link. This results in a bi-level programming problem which they solve iteratively as a sequence of linear programming problems by using sensitivity expressions for equilibrium assignment.

2.12 SPACE–TIME NETWORKS

So far, the time dimension has not been considered. In reality, flow moves at finite speeds through the network and conditions are not stationary. Most transport networks are subject to a morning and an afternoon peak as people travel to and from work. There are also longer term variations, for example flow tends to build up at certain locations during the summer vacation. In addition to this, there are irregular variations in flow due to things like occasional sporting events, accidents and road works.

The trajectory of a trip through space and time may appear something like Fig. 2.7. Clearly the timing of events, such as the arrival at bottleneck links or junctions, may affect parameters such as delay and capacity. One approach is to discretise time and include the time dimension in the network. This leads to a *space–time extended network* (STEN), where links are identified not just by the locations of their start and end nodes in space but also by their locations in time. By incorporating the time dimension into the network in this way, it is possible to apply some steady state network flow programming methods to dynamic situations, as will be shown later.

A STEN, plotted against space and time axes, is shown in Fig. 2.8. The vertical links represent delay. Two paths are shown. In the first, there is a delay of two intervals before entering link 3. The second path appears to overtake the first by entering link 2 later but entering link 3 earlier, constituting a violation of the first-in, first-out (FIFO) principle.

The STEN approach was originally proposed by Ford and Fulkerson (1962) for the study of the dynamic maximal flow problem, where each link has a capacity and a fixed traversal time. At each node, trip-makers can be either *forwarded* to the next link or *stored* for one period. This is a dynamic version of the steady state store-and-forward network. Zawack and Thompson (1987) adapted the STEN approach for dynamic traffic assignment on urban transportation networks with the terminology of green links for forwarding links and red links for the storage links. However, in his model only a many-to-one assignment problem is considered. More recently, Drissi-Kaitouni and Hameda-Benchekroun (1992) and Drissi-Kaitouni (1993) enhanced the STEN of Ford and Fulkerson (1962).

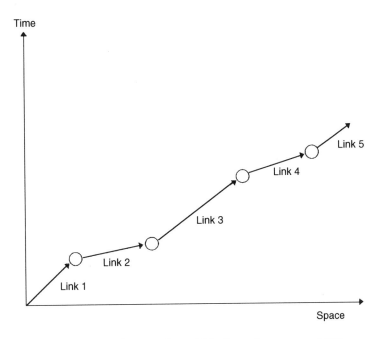

Figure 2.7 *Trajectory of a vehicle through space and time*

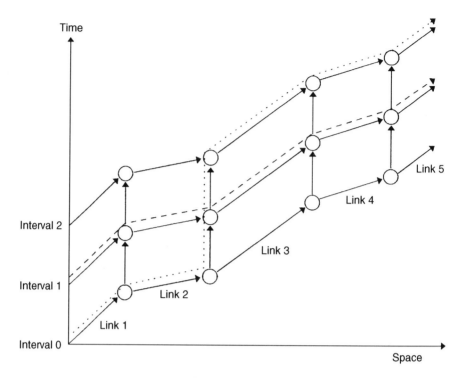

Figure 2.8 *An example STEN network with FIFO violation*

There are various ways in which the STEN may be generated. Following Drissi-Kaitouni (1993), a STEN may be constructed for a network $G(N, A)$ consisting of a set of nodes N and set of links A as follows:

- *Nodes*: Each node $n (n \in N)$ of the base network is expanded to $T + 1$ nodes $n(t)$, $t = 0, 1, \ldots, T$
- *Links*: For each link $a = (n, n') \in A$ and for each period $t = 0, 1, \ldots, T$ construct one node $l(t)$, one link $(n(t - \tau_a), l(t))$ if $t - \tau_a \geqslant 0$, one link $(l(t), n'(t))$ and one link $(l(t), l(t + 1))$ if $t + 1 \leqslant T$.

These three new links correspond to the fixed travel time τ_a on the original link a, the exit process and the queue at the given time period t respectively. Yang and Meng (1995) have extended the approach to allow an upper limit to be placed on the number of trips entering the network and to allow trip-makers to have a choice of departure time. The extension is as follows.

- *Origin node*: For each origin node o and for each period $t = 0, 1, \ldots, T$ construct two nodes $n(t)$ and $l(t)$, one link $(o, l(t))$, one link $(l(t), l(t + 1))$ if $t + 1 \leqslant T$, and one link $(l(t), n(t))$.

The first link represents the choice of departure time, the second link the wait at the origin o during period t, and the third link represents the departure process.

- *Destination nodes*: For each destination node d and for each period $t = 0, 1, \ldots, T$ construct one node $n(t)$ and one link $(n(t), d)$.

This link corresponds to the exit of traffic from the network at destination d during period t, where $t = 0, 1, \ldots, T$.

In general, one might expect traffic on links to conform, at least on average, to a FIFO rule. That is, after all, the way a queue behaves. With reference to Fig. 2.8, this means that one would expect the flow entering the path in interval 1 to pass the bottleneck on link 2 later and arrive at the destination later than flow entering the path in interval 0. It has been shown by Carey (1992) that FIFO leads to non-convex constraints on feasible link and path flows, making it difficult to impose in network flow programming methods.

2.13 NETWORK EQUILIBRIUM

As discussed in Chapter 1, the methods presented in this book make use of the concept of *network equilibrium*. Having defined the components of the network, it is now appropriate to discuss the concept of equilibrium in more detail. The discussion here on the stability of the traffic assignment model follows Watling (1996).

Consider any vector of path flows **h** (vector and matrix notation is introduced in Chapter 3, but for now a vector may be regarded as a list of variables, in this case of path flows). This vector is said to be *feasible* if it conforms to a prespecified trip table and prespecified link capacities (see Smith, 1987). The path flows h are said to be in *deterministic user equilibrium* if and only if they are feasible and

$h_r > 0$ implies $g_r(\mathbf{h}) \leqslant g_s(\mathbf{h})$ for all paths r and $s(r \neq s)$ connecting a given origin to a given destination.

That is, path r is used only if its cost is less than or equal to the cost of all other paths serving the same origin and destination. If the set of feasible path flows is non-empty, then the existence of equilibria can be guaranteed under very general conditions (see Chapter 5). Given the existence of equilibria, it is desirable to know something about their *sensitivity* and their *stability*. The subject of sensitivity is returned to frequently in subsequent chapters.

While it is realistic to expect trip-makers to *prefer* least cost paths, deterministic user equilibrium as defined above presupposes that all trip-makers perceive costs identically. In reality, costs are *perceived* differently by different trip-makers, leading to a spread of responses. This leads to the concept of *stochastic user equilibrium* whereby trip-makers choose options that they perceive to be least cost, but the perception of costs is subject to random variation (see Sheffi, 1985). Let $p_{rij}(\mathbf{g})$ denote the proportion of the total demand t_{ij} that would choose path r between origin i and destination j, expressed as a function of the vector of prevailing path costs **g**. Then the flow vector **h** is a stochastic user equilibrium if and only if

$$h_r = t_{ij} p_{rij}(\mathbf{g}(\mathbf{h})) \text{ for all paths connecting origin } i \text{ to destination } j$$

Conventionally, the path choice proportions for each origin i and destination j are assumed to follow a *discrete choice model*, such as the logit model. While discrete choice models are usually founded on the notion of randomly perceived costs, the proportions are generally treated as known functions of *fixed* costs and are therefore not themselves subject to statistical fluctuation. Hazelton *et al*. (1996) have recently criticised this approach as it does not allow for the feedback effect of the randomness in path choice on the path costs.

One attractive feature of stochastic user equilibrium is that the path flows are uniquely given (see Chapter 6), which is not generally the case under deterministic user equilibrium. As the amount of variation in the perception of the costs tends to zero, the stochastic user equilibrium assignment tends to a deterministic user equilibrium assignment. The existence and uniqueness of equilibria is dealt with in greater detail in Chapters 5 and 6.

The stability of a deterministic or stochastic user equilibrium depends on the dynamical response of the system under perturbation. When the system continually moves toward a new state where the perceived costs of trip-makers are reduced (if possible), and does so in a smooth way, then the system is said to be *stable*. In reality, individual trip-makers exhibit complex behaviour involving learning and habit, so stability cannot be presumed. However, it seems reasonable to assume that transportation systems hover near equilibrium, because the incentives generated by disequilibrium would, if decisions are taken rationally, tend to restore the system to a state of equilibrium. In a Darwinian world the costs of irrational behaviour are not sustainable in the longer term.

Let Δ_{rs} be a path swap vector with value -1 for element r, 1 for element s and 0 otherwise. Then the flow vector $\mathbf{h}$ is said to be *user optimised* if and only if

$h_r > 0$ implies $g_r(\mathbf{h}) \leqslant g_s(\mathbf{h} + \varepsilon \Delta_{rs})$ for $0 < \varepsilon \leqslant h_r$ and for all paths r and s connecting a given origin to a given destination.

Heydecker (1986) gives this definition the interpretation that 'flows are user optimised if any driver who changes to an alternative route will experience a cost which is at least as great as the old one on his old route'. In general, a user equilibrium need not be a user optimum, and indeed a user optimum need not be a user equilibrium. Consider the network shown in Fig. 2.9.

When the link cost functions are

$$c_1 = 2 + 2v_1$$

$$c_2 = 10 + v_2$$

$$c_3 = 2 + 2v_3$$

$$c_4 = 10 + v_4$$

$$c_5 = 2 + 2v_5$$

and the total flow from the origin to the destination is 10 units of flow (say, vehicles per second), the user equilibrium assignment occurs when $v_1 = v_5 = 5.143$ vehicles per second, $v_2 = v_4 = 4.857$ vehicles per second and $v_5 = 0.286$ vehicles per second, and all three paths between the origin and the destination are used. The trip cost by any path is 27.14 units of cost (say, dollars per trip). However, if link 3 were removed from the network, the user equilibrium link flows would be $v_1 = v_4 = v_2 = v_5 = 5$ vehicles per second and the trip cost would be 27 dollars per trip, a reduction of 0.14 dollars per trip. Note that the cost of the path using link 3 is now 26 dollars per trip, creating an incentive to use it. Thus in this example the user equilibrium ($v_1 = v_5 = 5.143$, $v_2 = v_4 = 4.857$ and $v_3 = 0.286$) is not a user optimum ($v_1 = v_4 = v_2 = v_5$ and $v_3 = 0$), and the user optimum is not a user equilibrium. This is known as Braess's paradox (see Braess, 1968, or Sheffi, 1985). The paradox arises since the addition of extra capacity in the form of link 3 has the effect of increasing the cost of a trip.

As a result of the possible discrepancy between the user equilibrium assignment and the user optimum assignment, and the concern that user equilibrium as defined earlier might •

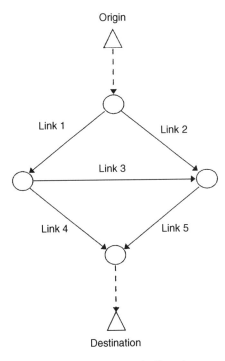

Origin

Link 1

Link 2

Link 3

Link 4

Link 5

Destination

Figure 2.9 *Braess's Paradox*

lack realism, Heydecker (1986) proposed a subtly different form of equilibrium, referred to here as a *Heydecker equilibrium*. The flow **h** is said to be a Heydecker equilibrium if and only if

$h_r > 0$ implies $g_r(\mathbf{h} + \varepsilon\Delta_{rs}) \leqslant g_s(\mathbf{h} + \varepsilon\Delta_{rs})$ for $0 < \varepsilon \leqslant h_r$ and for all paths r and s connecting a given origin to a given destination.

Here 'any driver who changes to an alternative route will experience a cost at least as great as the *new* one on his old route', and therefore have no incentive to return to his old route. It can be shown that a Heydecker equilibrium is also a user optimum.

The notion of equilibrium, when applied to dynamical systems, is intimately related to the concept of stability. A dynamical system is said to be stable if it tends to a *fixed point*, the equilibrium, after it has been perturbed. The conventional approach to the study of stability is to examine how the system behaves when perturbed. This leads Watling (1996) to the following definition of stability. A user equilibrium flow vector **h** is *locally stable* if and only if there exists $k_r (0 < k_r \leqslant h_r)$ such that

$h_r > 0$ implies $g_r(\mathbf{h} + \varepsilon\Delta_{rs}) \leqslant g_s(\mathbf{h} + \varepsilon\Delta_{rs})$ for $0 < \varepsilon \leqslant k_r < h_r$ and for all paths r and s connecting a given origin to a given destination.

In other words, a shift of flow *from* path r to path s should *encourage* the use of path r. Clearly a Heydecker equilibrium is by definition stable, but the deterministic user equilibrium may not be.

The concept of stability extends naturally to the stochastic user equilibrium situation. A stochastic user equilibrium flow vector **h** is said to be stable if and only if there exists

$k_r (0 < k_r \leqslant h_r)$ such that

$h_r > 0$ implies $t_{ij} p_{rij}(\mathbf{g}(\mathbf{h} + \varepsilon \Delta_{rs})) \geqslant h_r$ for $0 < \varepsilon \leqslant k_r < h_r$ and for all paths r connecting origin i to destination j.

Lack of model stability (as opposed to real world stability) can be attributable to non-separable, non-monotonically increasing link cost functions (see Watling, 1996). It is also possible that the equilibrium behaviour of a dynamical system is also dynamic. As an example, if choices made today depend on costs yesterday, the equilibrium may well not be a single fixed point but a number of fixed points.

2.14   CONCLUDING REMARKS

This chapter has concentrated on the components of transportation networks and the variables to be used in subsequent chapters. Two important building blocks of subsequent methods have also been presented. The first building block relates to the determination of shortest (or least cost) paths. Two shortest path algorithms are presented. The second building block relates to the determination of link choice proportions according to the logit path choice model. A matrix method that includes all paths is described.

In congested conditions it is necessary to introduce the time dimension to allow for temporary overloading. One approach is to extend the network to include the time dimension. There is, however, a difficulty associated with enforcing a first-in, first-out (FIFO) discipline in such extended networks. This difficulty is particularly pronounced for stochastic user equilibrium assignment, because all paths (including those implying unnecessary queuing) will have a non-zero probability of use. As a consequence, a different approach to time-dependency is adopted in subsequent chapters. Steady state (equilibrium) time slices linked by queues are considered. Queues are assumed to form only when capacity is reached and, when in existence, the queues have to be processed before the new arrivals. As argued later, this captures the more important features of FIFO.

This book is based on the assumption that real transportation systems tend to hover near an equilibrium and that therefore it is worthwhile exploring the equilibrium properties of transportation models. Although the stability of the transportation system cannot be presumed, the irrationality implied by decision-making that leads away from an equilibrium is not sustainable in the long term because of the cost to network users.

3
Optimality

3.1 INTRODUCTION

This chapter introduces basic mathematical optimality results made use of extensively in subsequent chapters. Many of the analytical methods presented in this book are best applied to large networks by formulating an *equivalent optimisation problem*, which may then be solved by efficient but standard optimisation techniques. For example, the assignment of traffic to paths according to the discrete choice logit model may be formulated as an optimisation problem, which may then be solved efficiently by an appropriate optimisation technique. The emphasis of the book is on the formulation and analysis of the equivalent optimisation problems rather than on their solution. In this chapter, the optimisation of a *convex objective function* subject to *linear constraints* is concentrated on, as this is the case that arises most frequently later. At the risk of losing some readers, use is made of vector and matrix notation because of the great economy this offers.

Following a description of matrix operations, there is a discussion of convexity relating both to the objective function and to the constraints. This is followed by a presentation of optimality conditions and a full justification for them. Dual variables are introduced via Farkas's lemma and interpreted in terms of sensitivity. The Lagrangian equation is described along with its properties, in particular the saddle point theorem. Expressions are derived for the sensitivity of the primal and dual variables to perturbations of the constraints. Variance and covariance expressions are obtained from the sensitivity expressions.

While some simple numerical optimisation methods are described in later chapters, this is more as a means to a solution rather as a central feature of the book. The view taken here is that, as computers increase in speed and memory, less emphasis will in any case be placed on issues of computational and memory efficiency (within limits, of course). Two numerical optimisation methods that are widely used in subsequent chapters are described in general terms here.

3.2 KINDS OF VARIABLE

A *variable* consists of one or more numbers whose values are yet to be determined. In this book, only two types of variable, namely *integers* and *reals*, will be encountered. When a single number is referred to, the variable is called a *scalar*. A scalar is indicated by a lower case character. When a list of numbers is referred to, the set of variables is called a *vector*. The numbers can be listed vertically as a column or horizontally as a row. Normally they are listed as a column, but may be *transposed* to form a row. A column vector is represented by a lower case character in bold and a transposition by a superscript T. For example, $\mathbf{x}^T = [x_1, x_2, \ldots, x_n]$ is a row vector of n variables $x_1, x_2, \ldots, x_n$.

Sometimes variables are cross-referenced to produce a table or *matrix*. A matrix is indicated by an upper case character in bold. For example,

$$\mathbf{X} = \begin{bmatrix} x_{11} & x_{12} & \cdots & x_{1n} \\ x_{21} & x_{22} & \cdots & x_{2n} \\ \cdots & & & \\ x_{m1} & x_{m2} & \cdots & x_{mn} \end{bmatrix}$$

is a matrix of m rows and n columns. Matrix $\mathbf{X}$ could represent a trip table, in which case element x_{ij} would represent the number of trips from an origin i to a destination j, there being m origins and n destinations (in the case of **trip tables**, it is conventional for the first subscript to represent the origin and the second subscript the destination). Note that

$$\mathbf{X}^T = \begin{bmatrix} x_{11} & x_{12} & \cdots & x_{m1} \\ x_{12} & x_{22} & \cdots & x_{m2} \\ \cdots & & & \\ x_{1n} & x_{2n} & \cdots & x_{mn} \end{bmatrix}$$

In many of the methods described later, however, a trip table is represented as a vector. This is done by putting the rows of $\mathbf{X}$ end to end, thus

$$\mathbf{t}^T = [x_{11} \quad x_{12} \quad \cdots \quad x_{1n} \quad x_{21} \quad x_{22} \quad \cdots \quad x_{2n} \quad \cdots \quad x_{m1} \quad x_{m2} \quad \cdots \quad x_{mn}]$$

When the cells of the trip table are numbered distinctly and contiguously, this may be rewritten as

$$\mathbf{t}^T = [t_1 \quad t_2 \quad \cdots \quad t_{mn}]$$

3.3 MATRIX OPERATIONS

When certain operations with vectors and matrices are defined, vector and matrix notation offers an economical language for expressing relationships, potentially involving a large number of variables.

If $\mathbf{X}$ and $\mathbf{Y}$ are two matrices with the same dimensions (the same number of rows and columns), then $\mathbf{Z} = \mathbf{X} + \mathbf{Y}$ is a new matrix having the same dimensions as $\mathbf{X}$ and $\mathbf{Y}$ with elements $z_{ij} = x_{ij} + y_{ij}$. The operation of addition is only defined if the two matrices have the same dimensions. This applies also to vectors (in fact, vectors may be regarded as matrices with only one column).

If $\mathbf{X}$ is a matrix with m rows and r columns, and $\mathbf{Y}$ a matrix with r rows and n columns, then $\mathbf{Z} = \mathbf{XY}$ is a new matrix with m rows and n columns having elements

$$z_{ij} = \sum_{k=1 \text{ to } r} x_{ik} y_{kj}.$$

where the sigma is a summation operator. The operation of multiplication is only defined if the number of columns of $\mathbf{X}$ is the same as the number of rows of $\mathbf{Y}$. The same compatibility requirement applies to the product of two vectors or of a vector by a matrix.

Provided the matrices are compatible, three or more matrices may be multiplied together. For example, $\mathbf{Z} = \mathbf{WXY}$ is a matrix with elements

$$z_{ij} = \sum_{l=1 \text{ to } m} w_{il} \sum_{k=1 \text{ to } r} x_{lk} y_{kj} = \sum_{l=1 \text{ to } m} \sum_{k=1 \text{ to } r} w_{il} x_{lk} y_{kj}.$$

The implication of the above relocation of the second sigma (the summation operator applying to subscript k) is that series of matrices may be multiplied together without storing intermediate matrices.

Note that the transpose of the product of two matrices is the product of the two matrices transposed, but *in reverse sequence*, as follows:

$$(\mathbf{AB})^{\mathrm{T}} = \mathbf{B}^{\mathrm{T}}\mathbf{A}^{\mathrm{T}}$$

Simultaneous equations may be economically written in vector and matrix notation. For example, the following pair of simultaneous equations in x_1 and x_2

$$\begin{bmatrix} b_1 \\ b_2 \end{bmatrix} = \begin{bmatrix} a_{11}x_1 + a_{12}x_2 \\ a_{21}x_1 + a_{22}x_2 \end{bmatrix} = \begin{bmatrix} a_{11} & a_{12} \\ a_{21} & a_{22} \end{bmatrix} \begin{bmatrix} x_1 \\ x_2 \end{bmatrix}$$

may be written as

$$\mathbf{b} = \mathbf{Ax}$$

Indeed, if the operation of **matrix inversion** is defined, the solution to the simultaneous equations may be expressed straightforwardly as

$$\mathbf{x} = \mathbf{A}^{-1}\mathbf{b}$$

For the simultaneous equations to have a unique solution for x, and therefore for $\mathbf{A}^{-1}$ to be unique, matrix $\mathbf{A}$ must be *non-singular*.

When $\mathbf{A}$ is as defined above, namely

$$\mathbf{A} = \begin{bmatrix} a_{11} & a_{12} \\ a_{21} & a_{22} \end{bmatrix}$$

then

$$\mathbf{A}^{-1} = \begin{bmatrix} a_{22}/(a_{11}a_{22} - a_{12}a_{21}) & -a_{12}/(a_{11}a_{22} - a_{12}a_{21}) \\ -a_{21}/(a_{11}a_{22} - a_{12}a_{21}) & a_{11}/(a_{11}a_{22} - a_{12}a_{21}) \end{bmatrix}$$

Hence

$$\begin{bmatrix} x_1 \\ x_2 \end{bmatrix} = \begin{bmatrix} a_{22}/(a_{11}a_{22} - a_{12}a_{21}) & -a_{12}/(a_{11}a_{22} - a_{12}a_{21}) \\ -a_{21}/(a_{11}a_{22} - a_{12}a_{21}) & a_{11}/(a_{11}a_{22} - a_{12}a_{21}) \end{bmatrix} \begin{bmatrix} b_1 \\ b_2 \end{bmatrix}$$

or

$$x_1 = b_1 a_{22}/(a_{11}a_{22} - a_{12}a_{21}) - b_2 a_{12}/(a_{11}a_{22} - a_{12}a_{21})$$

$$= (b_1 a_{22} - b_2 a_{12})/(a_{11}a_{22} - a_{12}a_{21})$$

$$x_2 = -b_1 a_{21}/(a_{11}a_{22} - a_{12}a_{21}) + b_2 a_{11}/(a_{11}a_{22} - a_{12}a_{21})$$

$$= (b_2 a_{11} - b_1 a_{21})/(a_{11}a_{22} - a_{12}a_{21})$$

Note that the elements of $\mathbf{A}^{-1}$ above are infinite if the *determinant* of $\mathbf{A}$ is zero, namely if $a_{11}a_{22} - a_{12}a_{21} = 0$. A non-singular matrix therefore has a non-zero determinant, and also a matrix with a non-zero determinant is non-singular. When $\mathbf{A}$ is non-singular, the simultaneous equations possess a unique solution for $\mathbf{x}$, namely that given above.

Furthermore, note that

$$\mathbf{A}\mathbf{A}^{-1} = \mathbf{A}^{-1}\mathbf{A} = \begin{bmatrix} 1 & 0 \\ 0 & 1 \end{bmatrix} = \mathbf{I}$$

where $\mathbf{I}$ is an *identity matrix*.

The product of any matrix $\mathbf{A}$ by an identity matrix $\mathbf{I}$ of appropriate dimension leaves the original matrix unchanged. Thus

$$\mathbf{A}\mathbf{I} = \mathbf{I}\mathbf{A} = \mathbf{A}$$

There are a number of properties that a non-singular matrix must possess. A set of simultaneous equations will not have a unique solution unless the number of linearly independent equations is equal to the number of unknowns, so a non-singular matrix must be square and all the rows must be linearly independent of one another. The *rank* of a matrix is the number of linearly independent rows. The number of linearly independent rows is equal to the number of linearly independent columns.

Note that the rank of the product of two matrices is equal to the lesser of the two ranks of the two matrices considered alone. Thus

$$\text{rank}(\mathbf{A}\mathbf{B}) = \text{Min}\{\text{rank}(\mathbf{A}), \text{rank}(\mathbf{B})\}$$

In some cases, a matrix conveniently decomposes into *blocks*, where the diagonal blocks are square. For example

$$\mathbf{A} = \begin{bmatrix} \mathbf{A}_{11} & \mathbf{A}_{12} \\ \mathbf{A}_{21} & \mathbf{A}_{22} \end{bmatrix}$$

In this case, the inverse may also be expressed in block form. Let $\mathbf{B} = \mathbf{A}^{-1}$, then

$$\mathbf{B} = \begin{bmatrix} \mathbf{A}_{11}^{-1} + \mathbf{A}_{11}^{-1}\mathbf{A}_{12}\mathbf{B}_{22}\mathbf{A}_{21}\mathbf{A}_{11}^{-1} & -\mathbf{A}_{11}^{-1}\mathbf{A}_{12}\mathbf{B}_{22} \\ -\mathbf{B}_{22}\mathbf{A}_{21}\mathbf{A}_{11}^{-1} & \mathbf{B}_{22} \end{bmatrix}$$

where

$$\mathbf{B}_{22} = (\mathbf{A}_{22} - \mathbf{A}_{21}\mathbf{A}_{11}^{-1}\mathbf{A}_{12})^{-1}$$

Such a block decomposition is most useful when one of the blocks, say $\mathbf{A}_{22}$, is zero. In this case

$$\mathbf{B}_{22} = -(\mathbf{A}_{21}\mathbf{A}_{11}^{-1}\mathbf{A}_{12})^{-1}$$

and

$$\mathbf{B} = \begin{bmatrix} \mathbf{A}_{11}^{-1} - \mathbf{A}_{11}^{-1}\mathbf{A}_{12}(\mathbf{A}_{21}\mathbf{A}_{11}^{-1}\mathbf{A}_{12})^{-1}\mathbf{A}_{21}\mathbf{A}_{11}^{-1} & \mathbf{A}_{11}^{-1}\mathbf{A}_{12}(\mathbf{A}_{21}\mathbf{A}_{11}^{-1}\mathbf{A}_{12})^{-1} \\ (\mathbf{A}_{21}\mathbf{A}_{11}^{-1}\mathbf{A}_{12})^{-1}\mathbf{A}_{21}\mathbf{A}_{11}^{-1} & -(\mathbf{A}_{21}\mathbf{A}_{11}^{-1}\mathbf{A}_{12})^{-1} \end{bmatrix}$$

Note that the inverse of the product of two matrices is the product of the inverse of the two square non-singular matrices, *in reverse order*:

$$(\mathbf{A}\mathbf{B})^{-1} = \mathbf{B}^{-1}\mathbf{A}^{-1}$$

The matrix $\mathbf{A}$ is *positive definite* when

$$\mathbf{x}^T\mathbf{A}\mathbf{x} > 0 \text{ for all vectors } \mathbf{x} \neq \mathbf{0}$$

where $\mathbf{0}$ is a vector all of whose elements are 0. Positive definite matrices are also non-singular.

Frequent use will be made subsequently of *functions* of vectors. These transform one or more variables into one or more other variables. If the function returns a single value, it is referred to as a *scalar-valued function*; if it returns more than one value, it is referred to as a *vector-valued function*. For example, the scalar-valued function

$$f(\mathbf{x}) = \mathbf{x}^T\mathbf{A}\mathbf{x}$$

returns a single positive value for any $\mathbf{x}$, provided $\mathbf{A}$ is positive definite.

If the function is differentiable, it may be differentiated partially with respect to each element of $\mathbf{x}$. The result is the *gradient* of $f(\mathbf{x})$. Define $\nabla f(\mathbf{x})$ as the gradient of function $f(\mathbf{x})$ at $\mathbf{x}$; ∇ is therefore an operator that performs the operation of partial differentiation on $f(\mathbf{x})$ with respect to each element of $\mathbf{x}$. For non-linear functions, the gradient will be a function of $\mathbf{x}$, so the value of $\mathbf{x}$ for which the gradient is evaluated must be specified. Thus $\nabla f(\mathbf{x})$ is the gradient of $f(\mathbf{x})$ *at* $\mathbf{x}$. The gradient is a row vector with as many elements as $\mathbf{x}$. For the above scalar-valued function

$$\nabla f(\mathbf{x}) = 2\mathbf{A}\mathbf{x}$$

Note that in this case, the gradients are themselves also a function of $\mathbf{x}$. When the operator is applied a second time to $f(\mathbf{x})$, the result is a matrix, namely in this case

$$\nabla^2 f(\mathbf{x}) = 2\mathbf{A}$$

This is referred to as the *Hessian* of $f(\mathbf{x})$.

3.4   OBJECTIVE FUNCTIONS

An *objective function* is a function that is optimised (either maximised or minimised). Consider the scalar-valued function

$$f(\mathbf{x}) = x_1(\ln(x_1) - 1) + x_2(\ln(x_2) - 1) \text{ with } x_1 > 0 \text{ and } x_2 > 0$$

This returns a single value for any given values of vector $\mathbf{x}$. This may be minimised with respect to $\mathbf{x}$. In this case, there are restrictions on the feasible values of $\mathbf{x}$. A useful property for an objective function to have, as will become clear later, is *convexity*.

A convex function is one for which

$$f(\mathbf{x}_1\lambda + \mathbf{x}_2(1 - \lambda)) \leqslant \lambda f(\mathbf{x}_1) + (1 - \lambda)f(\mathbf{x}_2) \text{ for } 1 \geqslant \lambda \geqslant 0$$

where $\mathbf{x}_1$ and $\mathbf{x}_2$ are any two values of $\mathbf{x}$ with $\mathbf{x}_1 \neq \mathbf{x}_2$. This suggests that if a straight line is drawn from $f(\mathbf{x}_1)$ to $f(\mathbf{x}_2)$, the function between $\mathbf{x}_1$ and $\mathbf{x}_2$ lies on or below the line connecting these two points. The function is *strictly convex* if

$$f(\mathbf{x}_1\lambda + \mathbf{x}_2(1 - \lambda)) < \lambda f(\mathbf{x}_1) + (1 - \lambda)f(\mathbf{x}_2) \text{ for } 1 > \lambda > 0$$

For a *continuously differentiable* function, convexity implies that

$$f(\mathbf{x} + \mathbf{h}) \geqslant f(\mathbf{x}) + \mathbf{h}^T \nabla f(\mathbf{x})$$

for any pairs of vectors $\mathbf{x}$ and $\mathbf{h}$. In geometric terms, this implies that the function value lies everywhere on or above a *hyperplane* drawn through the point $\mathbf{x}$ tangentially to $f(\mathbf{x})$.

For the function introduced earlier

$$\nabla f(\mathbf{x}) = [\ln(x_1), \ln(x_2)]^T$$

Note that the gradients are zero for $x_1 = x_2 = 1$, negative for values of x_1 and x_2 less than 1 and positive for values of x_1 and x_2 greater than 1. This suggests that $f(\mathbf{x})$ is convex and has a minimum at $x_1 = x_2 = 1$.

Define $\nabla^2 f(\mathbf{x})$ as the second derivative matrix, or *Hessian*, of $f(\mathbf{x})$ at $\mathbf{x}$; ∇^2 is therefore an operator that performs the operation of partial differentiation *twice*. The Hessian is a matrix with as many rows and columns as $\mathbf{x}$ has elements. The existence of the Hessian presupposes that $f(\mathbf{x})$ is twice differentiable at $\mathbf{x}$. Note that the Hessian matrix must be *symmetric* since the order of differentiation does not matter, namely

$$\partial^2 f(\mathbf{x})/\partial x_i \partial x_j = \partial^2 f(\mathbf{x})/\partial x_j \partial x_i$$

The first three terms of the *Taylor series expansion* of $f(\mathbf{x})$ provide a quadratic approximation to $f(\mathbf{x})$ at $\mathbf{x}$. For $\mathbf{h}$ with small values

$$f(\mathbf{x} + \mathbf{h}) \approx f(\mathbf{x}) + \mathbf{h}^T \nabla f(\mathbf{x}) + 0.5 \mathbf{h}^T \nabla^2 f(\mathbf{x}) \mathbf{h}$$

The goodness of the approximation depends on the closeness of $\mathbf{x} + \mathbf{h}$ to $\mathbf{x}$ and the closeness of $f(\mathbf{x})$ to a quadratic function. If $\nabla^2 f(\mathbf{x})$ is positive definite for some $\mathbf{x}$, then $f(\mathbf{x})$ is strictly convex in the region of $\mathbf{x}$. If $\nabla^2 f(\mathbf{x})$ is positive definite for any $\mathbf{x}$, then the function is everywhere strictly convex.

For the function introduced earlier

$$\nabla^2 f(\mathbf{x}) = \begin{bmatrix} 1/x_1 & 0 \\ 0 & 1/x_2 \end{bmatrix}$$

This is positive definite for all feasible values of x_1 and x_2. Hence $f(\mathbf{x})$ is everywhere strictly convex with, as noted earlier, a minimum at $x_1 = x_2 = 1$.

A convenient alternative definition of convexity is

$$(\mathbf{x} - \mathbf{y})^T (\nabla f(\mathbf{x}) - \nabla f(\mathbf{y})) \geqslant 0$$

This implies that if x_i is greater than y_i then the ith element of $\nabla f(\mathbf{x})$ is greater than the ith element of $\nabla f(\mathbf{y})$. For a strictly convex function

$$(\mathbf{x} - \mathbf{y})^T (\nabla f(\mathbf{x}) - \nabla f(\mathbf{y})) > 0 \text{ for } \mathbf{x} \neq \mathbf{y}$$

Figure 3.1 shows an example of a convex function $f(\mathbf{x})$. It resembles a bowl with a distinct lowest point at an objective function value of zero ($f = 0$). The contours of the bowl are indicated by the isoquants (lines defined by specific objective function values, for example $f = 15$). Note that the value of f along the line $A - B$ lies *below* the line.

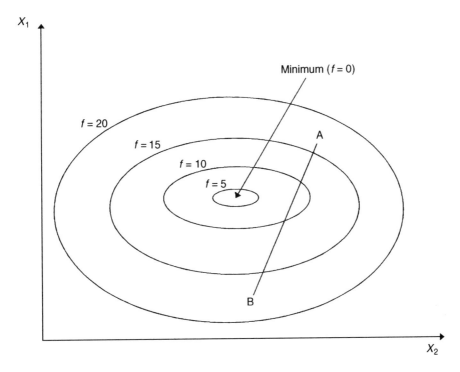

Figure 3.1 *An example convex function*

3.5 *CONSTRAINTS*

When optimising an objective function like $f(\mathbf{x})$, there are generally constraints placed on the values of $\mathbf{x}$. The example function presented in the previous section required that the values of $\mathbf{x}$ be positive. A particular value of $\mathbf{x}$ is said to be *feasible* if it satisfies the constraints. In this book, the constraints imposed on $\mathbf{x}$ are of linear form, namely

$$\mathbf{b} = \mathbf{Ax}$$

or

$$\mathbf{b} \geqslant \mathbf{Ax}$$

These constraints may be interpreted geometrically. In the former case, the set of feasible values is given by the intersection of m hyper planes, where m is the number of constraints or equivalently the number of elements in $\mathbf{b}$. In the latter case, the set of feasible values is given by the intersection of m *half spaces*, where as before m is the number of constraints. In general, both equality and inequality constraints are imposed in the problems encountered in subsequent chapters.

Many of the methods presented in this book require that the set of feasible values be *convex*, *bounded* and *compact*. Convexity implies that if $\mathbf{x}_1$ and $\mathbf{x}_2$ are feasible then so is $\mathbf{x}_1 \lambda + \mathbf{x}_2 (1 - \lambda)$ where $1 \geqslant \lambda \geqslant 0$. In geometric terms this implies that a line drawn between any two feasible points lies entirely within the set of feasible values. Fig. 3.2a shows a non-convex set (note that the line $A - B$, connecting feasible points A and B, does not lie entirely within the set) while Fig. 3.2b shows a convex set (note that it is

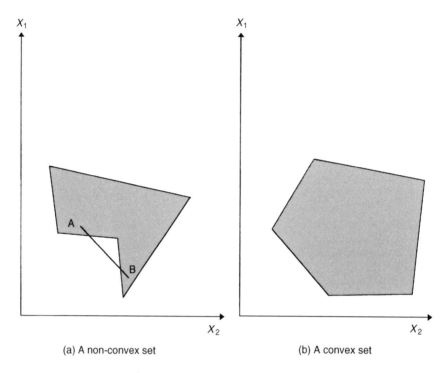

(a) A non-convex set   (b) A convex set

Figure 3.2 *Convex and non-convex sets*

impossible to draw a line A–B that *does not* lie entirely within the set). Boundedness implies that the boundaries of the set are themselves within the set (in other words feasible). Compactness means that there are no holes in the set of feasible values.

The linear inequality constraints above are convex because, for any two feasible points $\mathbf{x}_1$ and $\mathbf{x}_2$,

$$\mathbf{A}(\mathbf{x}_1\lambda + \mathbf{x}_2(1 - \lambda)) = \mathbf{A}\mathbf{x}_1\lambda + \mathbf{A}\mathbf{x}_2(1 - \lambda) \leqslant \mathbf{b}\lambda + \mathbf{b}(1 - \lambda) = \mathbf{b}$$

for $1 \geqslant \lambda \geqslant 0$. A similar argument applies for the linear equality constraints. In fact the linear equality constraints are simultaneously convex and concave.

Returning to the example presented earlier, the following constraints would ensure that $x_1 + x_2 = b_1$, that $x_1 \leqslant b_2$ and that $x_2 \leqslant b_3$.

$$\begin{bmatrix} b_1 \\ -b_1 \\ b_2 \\ b_3 \end{bmatrix} \geqslant \begin{bmatrix} 1 & 1 \\ -1 & -1 \\ 1 & 0 \\ 0 & 1 \end{bmatrix} \begin{bmatrix} x_1 \\ x_2 \end{bmatrix}$$

These constraints and the feasible set are illustrated in Fig. 3.3. Note that while any equality constraint may be represented by two inequality constraints, any inequality constraint may be represented by an equality constraint by introducing non-negative *slack variables*. This technique is useful in linear programming in particular.

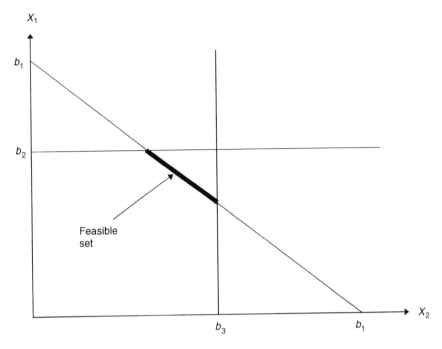

Figure 3.3 *Equality and inequality constraints*

3.6 *OPTIMALITY CONDITIONS*

Consider the problem

$$\text{Minimise } f(\mathbf{x}) \text{ with respect to } \mathbf{x} \text{ subject to } \mathbf{b} \geqslant \mathbf{Ax}$$

Suppose $\mathbf{x}^*$ is a local minimum to the above problem. Then a move away from $\mathbf{x}^*$ in any feasible direction should not decrease $f(\mathbf{x})$, at least locally, otherwise $\mathbf{x}^*$ is not a local minimum. Let $\mathbf{x}$ be any other feasible point. The convexity of the constraints implies that $\mathbf{x}' = \mathbf{x}\lambda + \mathbf{x}^*(1 - \lambda)$ is also a feasible point for $1 \geqslant \lambda \geqslant 0$. Thus

$$\mathbf{x}' - \mathbf{x}^* = \mathbf{x}\lambda + \mathbf{x}^*(1 - \lambda) - \mathbf{x}^* = (\mathbf{x} - \mathbf{x}^*)\lambda$$

is a step in a feasible direction. Then for a small step (λ small)

$$\lambda(\mathbf{x} - \mathbf{x}^*)^\mathrm{T}\nabla f(\mathbf{x}^*) \geqslant 0$$

Dividing both sides by $\lambda > 0$ yields the following *variational inequality*

$$(\mathbf{x} - \mathbf{x}^*)^\mathrm{T}\nabla \mathbf{f}(\mathbf{x}^*) \geqslant 0 \text{ for all feasible } \mathbf{x}$$

When $f(\mathbf{x})$ is convex and the constraint set is also convex, this variational inequality defines a global optimum solution.
If

$$(\mathbf{x} - \mathbf{x}^*)^\mathrm{T}\nabla f(\mathbf{x}^*) < 0 \text{ for some feasible } \mathbf{x}$$

then $(\mathbf{x} - \mathbf{x}^*)$ is a *descent direction*. Most optimisation methods for convex problems involve finding a descent direction and then moving in this direction until no more descent is achievable.

Note further that if $f(\mathbf{x})$ is strictly convex, the variational inequality implies that the optimum is not only global but also *unique*. Suppose otherwise, namely that there are two optima, $\mathbf{x}^*$ and $\mathbf{x}^{**}$, with $\mathbf{x}^* \neq \mathbf{x}^{**}$. Then

$$(\mathbf{x}^{**} - \mathbf{x}^*)^{\mathrm{T}} \nabla f(\mathbf{x}^*) \geqslant 0$$

and

$$(\mathbf{x}^* - \mathbf{x}^{**})^{\mathrm{T}} \nabla f(\mathbf{x}^{**}) \geqslant 0$$

Taken together these imply that

$$(\mathbf{x}^* - \mathbf{x}^{**})^{\mathrm{T}} (\nabla f(\mathbf{x}^*) - \nabla f(\mathbf{x}^{**})) \leqslant 0$$

which contradicts the strict convexity assumption, as this requires that

$$(\mathbf{x}^* - \mathbf{x}^{**})^{\mathrm{T}} (\nabla f(\mathbf{x}^*) - \nabla f(\mathbf{x}^{**})) > 0$$

(see Section 3.4).

If any constraints are *active* at the solution, a small step away from the solution in a feasible direction could lead to the relaxation of one or more active constraints. As before, the convexity of the constraints implies that, if $\mathbf{x}^*$ is the solution and $\mathbf{x}$ any other feasible point, then $\mathbf{x}' = \mathbf{x}\lambda + \mathbf{x}^*(1 - \lambda)$ is also a feasible point for $1 \geqslant \lambda \geqslant 0$. Thus for small λ, $\mathbf{x}' - \mathbf{x}^* = \mathbf{x}\lambda + \mathbf{x}^*(1 - \lambda) - \mathbf{x}^* = (\mathbf{x} - \mathbf{x}^*)\lambda$ is a small step in a feasible direction, so

$$\mathbf{A}(\mathbf{x}' - \mathbf{x}^*) = \mathbf{A}(\mathbf{x} - \mathbf{x}^*)\lambda \leqslant 0$$

the direction of the inequality being governed by the direction of the inequality of the original constraints. Dividing both sides by $\lambda > 0$ yields

$$\mathbf{A}(\mathbf{x} - \mathbf{x}^*) \leqslant 0$$

Where the constraints on $\mathbf{x}$ are simple non-negativity constraints of the form

$$\mathbf{x} \geqslant 0$$

the variational inequality simplifies to

$$\mathbf{x}^{*\mathrm{T}} \nabla f(\mathbf{x}^*) = 0 \text{ and } \mathbf{x}^{\mathrm{T}} \nabla f(\mathbf{x}^*) \geqslant 0 \text{ for all feasible } \mathbf{x}$$

To prove this, consider $\mathbf{x} = \mathbf{x}^*2$ and $\mathbf{x} = \mathbf{x}^*/2$, both of which must be feasible with respect to the non-negativity constraints as $\mathbf{x}^*$ is feasible. These two values of $\mathbf{x}$, when substituted into the variational inequality, yield

$$\mathbf{x}^{*\mathrm{T}} \nabla f(\mathbf{x}^*) \geqslant 0 \text{ and } \mathbf{x}^{*\mathrm{T}} \nabla f(\mathbf{x}^*) \leqslant 0$$

respectively. This gives the first part of the simplification. Reapplication of the variational inequality gives the second part. In fact, the second condition implies

$$\nabla f(\mathbf{x}^*) \geqslant 0$$

$$\mathbf{u}^* \geqslant \mathbf{0}$$

$$\mathbf{b} - \mathbf{Ax} \geqslant \mathbf{0}$$

which were the complementary slackness conditions obtained earlier.

Taken together, the saddle point theorem is obtained, which says that

$$L(\mathbf{x}, \mathbf{u}^*) \geqslant L(\mathbf{x}^*, \mathbf{u}^*) \geqslant L(\mathbf{x}^*, \mathbf{u})$$

and that

$$f(\mathbf{x}^*) = L(\mathbf{x}^*, \mathbf{u}^*)$$

At the optimum

$$\nabla_x L(\mathbf{x}^*, \mathbf{u}^*) = \mathbf{0}$$

$$\mathbf{u}^{*\mathrm{T}} \nabla_u L(\mathbf{x}^*, \mathbf{u}^*) = \mathbf{0}$$

$$\nabla_u L(\mathbf{x}^*, \mathbf{u}^*) \geqslant \mathbf{0}$$

$$\mathbf{u}^* \geqslant \mathbf{0}$$

where $\nabla_u L(\mathbf{x}^*, \mathbf{u}^*)$ is the gradient of $L(\mathbf{x}^*, \mathbf{u}^*)$ with respect to $\mathbf{u}$ given $\mathbf{x}^*$, evaluated at the optimum. These conditions correspond to the Kuhn–Tucker conditions. Given the appropriate convexity assumptions, these conditions are necessary and sufficient for an optimum.

In situations where $\mathbf{x}^*$ is expressible explicitly as a function of $\mathbf{u}$, as is the case with the entropy maximising models presented in this book, the saddle point theorem is of great practical importance.

In the example used in this chapter, $\mathbf{x}^*$ is a function of $\mathbf{u}$. To demonstrate this, consider the following Lagrangian equation:

$$L(\mathbf{x}, \mathbf{u}) = \sum_{i=1-2} x_i(\ln(x_i) - 1) - \mathbf{u}^{\mathrm{T}}(\mathbf{b} - \mathbf{Ax})$$

For given $\mathbf{u}$ the optimum $\mathbf{x}$ is given by

$$\nabla_x L(\mathbf{x}^*, \mathbf{u}) = \ln(\mathbf{x}^*) + \mathbf{A}^{\mathrm{T}}\mathbf{u} = \mathbf{0}$$

Hence

$$x_i^* = \exp\left(-\sum_{j=1-4} a_{ji}u_j\right) \qquad \text{for } i = 1 \text{ to } 2$$

An efficient way to obtain the optimal value of $\mathbf{x}$ is modify the values of $\mathbf{u}$ iteratively until the complementary slackness conditions:

$$\mathbf{u}^{*\mathrm{T}}(\mathbf{b} - \mathbf{Ax}^*(\mathbf{u}^*)) = \mathbf{0}$$

$$\mathbf{u}^* \geqslant \mathbf{0}$$

$$\mathbf{b} - \mathbf{Ax}^*(\mathbf{u}^*) \geqslant \mathbf{0}$$

obtain, where $\mathbf{x}^*(\mathbf{u})$ is the optimal value of $\mathbf{x}$ given $\mathbf{u}$, and by extension, $\mathbf{x}^*(\mathbf{u}^*)$ is the optimal value of $\mathbf{x}$ given the optimal value of $\mathbf{u}$ (the saddle point). The values of $\mathbf{u}^*$ will be unique if the constraints are linearly independent, which is the case here as none of the constraints are actually redundant. The value of $u_j{}^*$ will only be positive if constraint j is active.

3.9 SENSITIVITIES

In Section 3.7, the interpretation of dual variables as sensitivity measures was indicated. In the subsequent chapters, the constraints themselves are sometimes not known with certainty. It is therefore useful to be able to examine the sensitivity of the optimal primal variables $\mathbf{x}^*$ and optimal dual variables $\mathbf{u}^*$ to the constraints $\mathbf{b}$. The following derivation follows Bell (1985).

Consider the case where all constraints are equality ones and where the optimal primal variables $\mathbf{x}^*$ are functions of the dual variables $\mathbf{u}$. Then

$$\mathbf{b} = \mathbf{A}\mathbf{x}^*(\mathbf{u})$$

The relationship between a small change in $\mathbf{b}$ and a small change in $\mathbf{u}$ is given by

$$\Delta\mathbf{b} = \mathbf{A}\mathbf{x}^*(\Delta\mathbf{u})$$

Let $\mathbf{J}(\mathbf{u}) = \nabla\mathbf{x}^*(\mathbf{u})$ be the matrix of partial differentials of all elements of $\mathbf{x}^*$ with respect to all elements of $\mathbf{u}$, referred to as the *Jacobian* of $\mathbf{x}^*(\mathbf{u})$. As shown, this is a function of $\mathbf{u}$ itself, but for simplicity the Jacobian will be written as $\mathbf{J}$. Thus $\mathbf{J}$ is the sensitivity of $\mathbf{x}^*$ to $\mathbf{u}$, so for small perturbations

$$\Delta\mathbf{b} = \mathbf{A}\Delta\mathbf{x}^* \approx \mathbf{A}\mathbf{J}\Delta\mathbf{u}$$

If the constraints are linearly independent, $\mathbf{A}$ has full row rank, so $\mathbf{A}\mathbf{J}$ is non-singular and invertible. Therefore $(\mathbf{A}\mathbf{J})^{-1}$ gives the sensitivity of $\mathbf{u}$ to small changes in $\mathbf{b}$, and

$$\Delta\mathbf{u} \approx (\mathbf{A}\mathbf{J})^{-1}\Delta\mathbf{b}$$

Since the sensitivity of $\mathbf{x}^*$ to $\mathbf{u}$ is given by $\mathbf{J}$, the sensitivity of $\mathbf{x}^*$ to $\mathbf{b}$ is given by $\mathbf{J}(\mathbf{A}\mathbf{J})^{-1}$. Hence for small perturbations

$$\Delta\mathbf{x}^* \approx \mathbf{J}(\mathbf{A}\mathbf{J})^{-1}\Delta\mathbf{b}$$

These sensitivity expressions may be used to get approximate variances and covariances for the fitted values $\mathbf{x}^*$. Define $\mathbf{V}(\mathbf{b})$ as the variance–covariance matrix of $\mathbf{b}$, namely

$$\mathbf{V}(\mathbf{b}) = E\{\Delta\mathbf{b}\Delta\mathbf{b}^{\mathsf{T}}\}$$

where $E\{.\}$ is the expected value of $\{.\}$. The diagonal elements of this matrix correspond to the variances of $\mathbf{b}$ and the off-diagonal elements to the covariances of $\mathbf{b}$. It follows from the definition of the variance–covariance matrix that

$$\mathbf{V}(\mathbf{u}) \approx (\mathbf{A}\mathbf{J})^{-1}\mathbf{V}(\mathbf{b})(\mathbf{J}^{\mathsf{T}}\mathbf{A}^{\mathsf{T}})^{-1}$$

This provides the variances and covariances of the dual variables as functions of the variances and covariances of the constraints, approximately.

Furthermore
$$V(\mathbf{x}^*) \approx \mathbf{J}V(\mathbf{u})\mathbf{J}^T \approx \mathbf{J}(\mathbf{AJ})^{-1}V(\mathbf{b})(\mathbf{J}^T\mathbf{A}^T)^{-1}\mathbf{J}^T$$

The variances and covariances of the primal variables are now expressed as a function of the variances and covariances of the constraints, via the variances and covariances for the dual variables.

In the case of the example presented earlier

$$x_i^* = \exp\left(-\sum_{j=1-4} a_{ji}u_j\right) \qquad \text{for} \quad i = 1 \text{ to } 2$$

which implies that

$$dx_i^*/du_j = -a_{ji}x_i^*$$

If $\mathbf{D}$ is a diagonal matrix with elements x_i^* on the principal diagonal and zeros elsewhere (namely if $\mathbf{D}$ has elements $d_{ii} = x_i^*$ for all i and $d_{ij} = 0$ for all i and j where $i \neq j$), then

$$\mathbf{J} = -\mathbf{D}\mathbf{A}^T$$

It will be recalled from Fig. 3.4 that, since $b_1 > 2$, there is pressure on the second constraint only, namely

$$b_1 = x_1 + x_2$$

with $u_2 > 0$ and $u_1 = u_3 = u_4 = 0$. The other constraints are inactive at the solution.

After removal of the three inactive constraints, the Jacobian becomes

$$\mathbf{J} = -\mathbf{D}\mathbf{A}_a^T = -\begin{bmatrix} x_1^* & 0 \\ 0 & x_2^* \end{bmatrix}\begin{bmatrix} -1 \\ -1 \end{bmatrix} = \begin{bmatrix} x_1^* \\ x_2^* \end{bmatrix}$$

so

$$\mathbf{A}_a\mathbf{J} = -\mathbf{A}_a\mathbf{D}\mathbf{A}_a^T = -(x_1^* + x_2^*)$$

Let $V(b_1)$ be the variance of the one active constraint. Then the variance of the one non-zero dual variable is

$$V(u_2) \approx V(b_1)/(x_1^* + x_2^*)^2 = V(b_1)/b_1^2$$

and

$$V(\mathbf{x}^*) \approx \begin{bmatrix} x_1^{*2}V(b_1)/b_1^2 & x_1^*x_2^*V(b_1)/b_1^2 \\ x_1^*x_2^*V(b_1)/b_1^2 & x_2^{*2}V(b_1)/b_1^2 \end{bmatrix}$$

Given a mean and variance for b_1 of, say, 10 vehicles per second and 10 vehicles per second squared respectively, the mean values for both x_1^* and x_2^* would be 5 vehicles per second as a result of the symmetry of the problem. The variances for both x_1^* and x_2^*, as well as their covariance, would be 2.5 vehicles per second squared. Therefore an approximate 95% confidence interval for both x_1^* and x_2^* would be approximately ±3.1 vehicles per second.

Actually these numerical results follow directly from the symmetry of the example. When b_1 increases by ∇b_1, x_1^* and x_2^* increase in tandem by $0.5\Delta b_1$. The variances of x_1^* and x_2^* are therefore both one quarter of that for b_1. Since

$$V(b_1) = V(x_1^*) + V(x_2^*) + 2C(x_1^*, x_2^*),$$

where $C(x_1{}^*, x_2{}^*)$ is the covariance between $x_1{}^*$ and $x_2{}^*$, it follows that the covariance is also one quarter of the variance of b_1.

Where the primal variables are not an explicit function of the dual variables, approximate sensitivity expressions for the primal and dual variables may nonetheless be obtained. Consider the problem

$$\text{Minimise } f(\mathbf{x}) \text{ subject to } \mathbf{b} = \mathbf{Ax}$$

At the optimum

$$\nabla f(\mathbf{x}^*) = -\mathbf{A}^\mathrm{T} \mathbf{u}$$

Differentiating both sides by $\mathbf{u}$ yields

$$\nabla^2 f(\mathbf{x}^*)(d\mathbf{x}^*/d\mathbf{u}) = -\mathbf{A}^\mathrm{T}$$

If $f(\mathbf{x}^*)$ is convex then $\nabla^2 f(\mathbf{x}^*)$ is positive definite and invertible, so

$$(d\mathbf{x}^*/d\mathbf{u}) = -(\nabla^2 f(\mathbf{x}^*))^{-1}\mathbf{A}^\mathrm{T}$$

The sensitivities sought are the effect of a small change in the constraints on the optimal primal variables, namely $(d\mathbf{x}^*/d\mathbf{b})$. Using the chain rule of differentiation and then by substitution

$$(d\mathbf{x}^*/d\mathbf{b}) = (d\mathbf{x}^*/d\mathbf{u})/(d\mathbf{b}/d\mathbf{u}) = -(\nabla^2 f(\mathbf{x}^*))^{-1}\mathbf{A}^\mathrm{T}/(d\mathbf{b}/d\mathbf{u})$$

Note that by differentiating the constraints with respect to the primal variables $\mathbf{x}^*$ the following is obtained:

$$d\mathbf{b}/d\mathbf{x}^* = \mathbf{A}$$

By applying the chain rule of differentiation again,

$$(d\mathbf{b}/d\mathbf{u}) = (d\mathbf{b}/d\mathbf{x}^*)(d\mathbf{x}^*/d\mathbf{u}) = -\mathbf{A}(\nabla^2 f(\mathbf{x}^*))^{-1}\mathbf{A}^\mathrm{T}$$

If the constraints are linearly independent, $(d\mathbf{b}/d\mathbf{u})$ will be non-singular and invertible. Hence

$$(d\mathbf{u}/d\mathbf{b}) = -(\mathbf{A}(\nabla^2 f(\mathbf{x}^*))^{-1}\mathbf{A}^\mathrm{T})^{-1}$$

Finally

$$(d\mathbf{x}^*/d\mathbf{b}) = (d\mathbf{x}^*/d\mathbf{u})(d\mathbf{u}/d\mathbf{b}) = (\nabla^2 f(\mathbf{x}^*))^{-1}\mathbf{A}^\mathrm{T}(\mathbf{A}(\nabla^2 f(\mathbf{x}^*))^{-1}\mathbf{A}^\mathrm{T})^{-1}$$

The approximate variance–covariance expressions are thus

$$V(\mathbf{u}^*) \approx (d\mathbf{u}^*/d\mathbf{b})V(\mathbf{b})(d\mathbf{u}^*/d\mathbf{b})^\mathrm{T}$$

and

$$V(\mathbf{x}^*) \approx (d\mathbf{x}^*/d\mathbf{b})V(\mathbf{b})(d\mathbf{x}^*/d\mathbf{b})^\mathrm{T}$$

In the numerical example given earlier

$$(\nabla^2 f(\mathbf{x}^*))^{-1} = \begin{bmatrix} x_1{}^* & 0 \\ 0 & x_2{}^* \end{bmatrix} = \mathbf{D}$$

so in fact these expressions encompass the earlier ones.

3.10 METHOD OF SUCCESSIVE AVERAGES

The optimisation problem

$$\text{Minimise } f(\mathbf{x}) \text{ subject to } \mathbf{b} = \mathbf{A}\mathbf{x}$$

can generally be solved iteratively by the method of successive averages. At any point $\mathbf{x}$, the negative of the gradient, namely $-\nabla f(\mathbf{x})$, provides a *descent direction*. The maximum feasible descent from a point $\mathbf{x}'$ is found by solving the following linear programming problem:

$$\text{Minimise } \mathbf{x}^{\mathrm{T}}\nabla f(\mathbf{x}') \text{ subject to } \mathbf{b} = \mathbf{A}\mathbf{x}$$

This problem is referred to as an *auxiliary problem*, and its solution, call it $\mathbf{x}^*$, is an *auxiliary point*. The method of successive averages takes the average of a sequence of auxiliary points, where each auxiliary point is the solution of an auxiliary problem based on the previous average of auxiliary points. Hence if $\mathbf{x}_n{}^*$ is the nth auxiliary point, then on iteration N

$$\mathbf{x}' \leftarrow (1/N) \sum_{n=1 \text{ to } N} \mathbf{x}_n{}^*$$

But

$$(1/N) \sum_{n=1 \text{ to } N} \mathbf{x}_n{}^* = (1/N)\mathbf{x}_N{}^* + (1/N) \sum_{n=1 \text{ to } N-1} \mathbf{x}_n{}^*$$

hence

$$\mathbf{x}'_{\text{new}} \leftarrow (1/N)\mathbf{x}_N{}^* + (1 - 1/N)\mathbf{x}'_{\text{old}}$$

where $\mathbf{x}_N{}^*$ is found by solving the above auxiliary problem. The convergence of the method may be slow, but is assured for convex optimisation problems.

3.11 ITERATIVE BALANCING

Optimisation problems associated with the notion of *entropy maximisation* arise frequently in subsequent chapters. These problems have the following general form:

$$\text{Minimise } \mathbf{x}^{\mathrm{T}}(\ln(\mathbf{x}) - \mathbf{1}) \text{ subject to } \mathbf{b} = \mathbf{A}\mathbf{x}$$

where $\mathbf{1}$ is a vector of appropriate dimension all of whose elements are one. When $\mathbf{A}$ is a matrix with elements that are both non-negative and less than or equal to one ($1 \geqslant a_{ij} \geqslant 0$), *iterative balancing* may be used to find both primal and dual variables efficiently. The numerical example adopted earlier in this chapter has exactly this form.
    The Lagrangian equation is

$$L(\mathbf{x}, \mathbf{u}) = \mathbf{x}^{\mathrm{T}}(\ln(\mathbf{x}) - \mathbf{1}) - \mathbf{u}^{\mathrm{T}}(\mathbf{b} - \mathbf{A}\mathbf{x})$$

For given $\mathbf{u}$ the optimum $\mathbf{x}$ is given by

$$\nabla_x L(\mathbf{x}^*, \mathbf{u}) = \ln(\mathbf{x}^*) + \mathbf{A}^{\mathrm{T}}\mathbf{u} = \mathbf{0}$$

which allows $\mathbf{x}^*$ to be expressed as a function of the dual variables as follows:

$$\mathbf{x}^* = \exp(-\mathbf{A}^T\mathbf{u})$$

The following iterative algorithm calculates $\mathbf{u}^*$ and $\mathbf{x}^*$.

Iterative balancing algorithm for equality constraints
Step 1   (initialisation)
         $\mathbf{u} \leftarrow \mathbf{0}$

Step 2   (iterative balancing)
         repeat
                 for all constraints j
                     $\mathbf{x}^* \leftarrow \exp(-\mathbf{A}^T\mathbf{u})$
                     if $b_j \neq \mathbf{a}_j^T\mathbf{x}^*$ then $u_j \leftarrow u_j - \ln(b_j) + \ln(\mathbf{a}_j^T\mathbf{x}^*)$
         until convergence.

The algorithm has converged when the elements of $\mathbf{u}$ cease to change very much between iterations or when the constraints are satisfied sufficiently. Convergence of the algorithm may be shown as follows. Note that the effect on $L(\mathbf{x}^*, \mathbf{u})$ of changing $\mathbf{u}$ is

$$dL(\mathbf{x}^*, \mathbf{u})/d\mathbf{u} = \nabla_x L(\mathbf{x}^*, \mathbf{u})(\partial\mathbf{x}^*/\partial\mathbf{u}) - (\mathbf{b} - \mathbf{A}\mathbf{x}^*) = -(\mathbf{b} - \mathbf{A}\mathbf{x}^*)$$

since

$$\nabla_x L(\mathbf{x}^*, \mathbf{u}) = \mathbf{0}^T$$

Note also that the dual variables at the optimum satisfy

$$L(\mathbf{x}^*, \mathbf{u}) \leqslant L(\mathbf{x}^*, \mathbf{u}^*)$$

so the intention of the algorithm is to maximise $L(\mathbf{x}^*, \mathbf{u})$ with respect to $\mathbf{u}$. When $b_j > \mathbf{a}_j^T\mathbf{x}^*$, $dL(\mathbf{x}^*, \mathbf{u})/du_i < 0$, so u_i should be decreased. This will cause the elements of $\mathbf{x}^*$ to increase, because all the elements of $\mathbf{a}_j$ are non-negative. Note that $dL(\mathbf{x}^*, \mathbf{u})/du_i$ is negative so long as $b_j > \mathbf{a}_j^T\mathbf{x}^*$. Therefore, to increase $L(\mathbf{x}^*, \mathbf{u})$ as much as possible, u_j should be decreased until $b_j = \mathbf{a}_j^T\mathbf{x}^*$, but not beyond. The change to u_j made in Step 2 of the iterative balancing algorithm ensures that $b_j \geqslant \mathbf{a}_j^T\mathbf{x}^*$ after the change, since $1 \geqslant a_{ij} \geqslant 0$. Conversely, when $b_j < \mathbf{a}_j^T\mathbf{x}^*$, $dL(\mathbf{x}^*, \mathbf{u})/du_i > 0$, so u_i should be increased. This will cause the elements of $\mathbf{x}^*$ to decrease, because all the elements of $\mathbf{a}_j$ are non-negative. Note that $dL(\mathbf{x}^*, \mathbf{u})/du_i$ is positive so long as $b_j > \mathbf{a}_j^T\mathbf{x}^*$. Therefore, to increase $L(\mathbf{x}^*, \mathbf{u})$ as much as possible, u_j should be increased until $b_j = \mathbf{a}_j^T\mathbf{x}^*$, but not beyond. Step 2 of the algorithm does this also. As $L(\mathbf{x}^*, \mathbf{u})$ is always increasing the algorithm converges. After convergence $L(\mathbf{x}^*, \mathbf{u}) = L(\mathbf{x}^*, \mathbf{u}^*)$ and $\mathbf{u} = \mathbf{u}^*$.

If the problem has instead inequality constraints, say

$$\text{Minimise } \mathbf{x}^T(\ln(\mathbf{x}) - 1) \text{ subject to } \mathbf{b} \leqslant \mathbf{A}\mathbf{x}$$

the algorithm may be modified as follows:

Iterative balancing algorithm for inequality constraints
Step 1   (initialisation)
         $\mathbf{u} \leftarrow \mathbf{0}$

Step 2   (iterative balancing)
```
      repeat
              for all constraints j
                   x* ← exp(−Aᵀu)
                   if bⱼ > aⱼᵀx* then uⱼ ← uⱼ − ln(bⱼ) + ln(aⱼᵀx*)
      until convergence.
```

The modifications to $\mathbf{u}$ are asymmetrical and result in $\mathbf{u} \leqslant \mathbf{0}$. The proof of convergence is as before only now there is only the first case to consider, since, when $b_j \leqslant \mathbf{a}_j^T\mathbf{x}^*$, constraint j is inactive and u_j remains equal to 0. When the direction of the inequality in the constraint is reversed, the direction of the inequality in the algorithm is also reversed.

Iterative balancing is easy to implement and delivers useful information when there are minor inconsistencies in the constraints. If any elements of $\mathbf{u}$ are tending to either plus or minus infinity, the corresponding constraints may be checked for consistency. The algorithm may of course be extended to handle a mixture of equality and inequality constraints. Redundant constraints do not have to be removed beforehand. If the same constraint is present twice, the second one will appear to be inactive whether or not the first one is active. The method of successive averages is more widely applicable but converges more slowly than iterative balancing, so, where applicable, iterative balancing is to be preferred.

3.12 SUMMARY

Extensive use is made of matrix notation in subsequent chapters. Table 3.1 lists useful matrix operations, Table 3.2 lists useful matrix properties and Table 3.3 lists useful matrix types.

For the optimisation of a convex objective function $f(\mathbf{x})$ subject to linear constraints, the optimality, complementary slackness and sensitivity conditions shown in Table 3.4 apply. Point $\mathbf{x}$ is any feasible point and point $\mathbf{x}^*$ is the optimal solution to the problem. The optimality conditions provide variational inequalities which define $\mathbf{x}^*$. The optimal dual variable values are given by the complementary slackness conditions. The dual variables will only be uniquely defined if the constraints are linearly independent. The gradients of the objective function at the solution are given by the dual variables.

When both the objective function and the constraints are convex, the variational inequalities define a unique, global optimum.

Where the optimal primal variables $\mathbf{x}^*$ can be expressed explicitly as a function of the dual variables $\mathbf{u}$, as is often the case in the subsequent chapters, it is generally convenient to use the saddle point theorem to calculate the optimal dual variable values en route to calculating the optimal fitted values. Provided the elements of the constraint matrix $\mathbf{A}$ lie in the range 0 to 1 inclusive, the dual variable values may be calculated by iterative balancing.

Linear sensitivity expressions are derived for the primal and dual variables with respect to small changes of the constraints. From these sensitivity expressions, approximate expressions for the variances and covariances of both the primal and dual variables, in terms of the variances and covariances of the active constraints, may be obtained. Where $\mathbf{x}^*$ is an explicit function of $\mathbf{u}$, the Jacobian of $\mathbf{x}^*(\mathbf{u})$ is required. Where this is not the case, $(\nabla^2 f(\mathbf{x}^*))^{-1}$ is required.

Table 3.1 *Summary of matrix operations*

Operation	Description
Transposition	A matrix is transposed when the rows become the columns (the first row becomes the first column, etc.). As a result the columns become the rows. Transposition is denoted by a superscript T. Note that $(\mathbf{A}^T)^T = \mathbf{A}$.
Addition/subtraction	Two (or more) matrices may be added/subtracted if they are compatible. In this case, compatibility means that they have the same number of rows and columns. Note that $\mathbf{A} + \mathbf{B} = \mathbf{B} + \mathbf{A}$.
Multiplication	The *first* element of the first row of the first matrix is multiplied by the *first* element of the first column of the second matrix, and added to the *second* element of the first row of the first matrix multiplied by the second element of the first column of the second matrix, etc. The result, when this is repeated for *all* the elements of the first row of the first matrix and those of the first column of the second matrix, is the element of the first row and first column of the product matrix. Repeating the above procedure for the *first* row of the first matrix and the *second* column of the second matrix gives the element of the first row and second column of the product matrix, etc. The product is only defined for compatible matrices, which in this case means that the first matrix must have the same number of columns as the second matrix has rows. The product matrix has the same number of rows as the first matrix and the same number of columns as the second matrix. Note that the sequence of multiplication matters (i.e. $\mathbf{AB} \neq \mathbf{BA}$, in general).
Inverse	When a matrix is pre- or postmultiplied by its inverse (if it exists) the result is the identity matrix (i.e. $\mathbf{AA}^{-1} = \mathbf{A}^{-1}\mathbf{A} = \mathbf{I}$). If the inverse of a matrix exists, the matrix is non-singular.
Partial differentiation	The operator ∇ denotes partial differentiation. $\nabla f(\mathbf{x})$ is a column vector of the partial derivatives of $f(\mathbf{x})$ evaluated at $\mathbf{x}$. Where the function inputs more than one variable, and the partial differentiation does not apply to all, the variables included in the differentiation are shown as subscripts. Thus $\nabla_x L(\mathbf{x}^*, \mathbf{u})$ is the partial derivative of $L(\mathbf{x}^*, \mathbf{u})$ with respect to $\mathbf{x}$ evaluated at $\mathbf{x}^*$ and $\mathbf{u}$.

Table 3.2 *Matrix properties*

Property	Description
Rank	The number of linearly independent rows, which is equal to the number of linearly independent columns. The rank of a matrix can be no more than the number of rows or columns, whichever is less. The rank of the product of two matrices is equal to the lesser of the ranks of the two matrices before multiplication.
Non-singular	A matrix is non-singular if it is square and has maximal rank. If a matrix is non-singular then its inverse exists.
Positive definite	A matrix is positive definite if, when pre- and post-multiplied by the same non-zero vector, the result is always a positive scalar (i.e. $\mathbf{A}$ is positive definite if and only if $\mathbf{x}^T \mathbf{A} \mathbf{x} > 0$ for any $\mathbf{x} \neq \mathbf{0}$). Diagonal matrices are positive definite provided all the elements on the principal diagonal are positive. Variance–covariance matrices are always positive definite.

Table 3.3   *Special matrix types*

Type	Description
Vector	A matrix consisting of one column only. Special vectors encountered in subsequent chapters are **1** which is a vector all of whose elements are 1, and **0** which is a vector all of whose elements are 0.
Symmetrical	A symmetrical matrix is unchanged by transposition ($\mathbf{A} = \mathbf{A}^\mathsf{T}$)
Diagonal	A matrix with non-zero entries on the principal diagonal (the cells from top left to bottom right of the matrix) and zeros elsewhere. It is square and non-singular. When the diagonal elements are positive it is also positive definite.
Identity	A diagonal matrix with the value 1 entered on the principal diagonal, conventionally written **I**. Any matrix multiplied by the identity matrix is left unchanged.
Jacobian	A matrix of first order partial derivatives or gradients. When a vector-valued function of a vector, for example costs as a function of flows $\mathbf{c}(\mathbf{v})$, is differentiated partially the result is the Jacobian of the cost function, written $\nabla\mathbf{c}(\mathbf{v})$. The vector in brackets denotes both the variable of differentiation and the point at which the partial derivatives are evaluated.
Hessian	A matrix of second order partial derivatives. When a scalar-valued function of a vector, for example an objective function $f(\mathbf{x})$, is differentiated partially twice, the result is the Hessian of the objective function, written $\nabla^2 f(\mathbf{x})$. The vector in brackets denotes both the variable of differentiation and the point at which the partial derivatives are evaluated. The Hessian is symmetrical. If $f(\mathbf{x})$ is a convex function then its Hessian is positive definite.

Table 3.4   *Optimality, complementary slackness and sensitivity conditions*

Constraints	$\mathbf{b} \geqslant \mathbf{Ax}$	$\mathbf{b} = \mathbf{Ax}$
Variational (in)equalities which define an optimum	$(\mathbf{x} - \mathbf{x}^*)^\mathsf{T}\nabla f(\mathbf{x}^*) \geqslant 0$ $\mathbf{A}(\mathbf{x} - \mathbf{x}^*) \leqslant \mathbf{0}$ for all feasible $\mathbf{x}$	$(\mathbf{x} - \mathbf{x}^*)^\mathsf{T}\nabla f(\mathbf{x}^*) = 0$ $\mathbf{A}(\mathbf{x} - \mathbf{x}^*) = \mathbf{0}$ for all feasible $\mathbf{x}$
Complementary slackness conditions that define the dual variable values	$\mathbf{u}^{*\mathsf{T}}(\mathbf{b} - \mathbf{Ax}^*) = 0$ $\mathbf{u}^* \geqslant \mathbf{0}$ $\mathbf{b} - \mathbf{Ax}^* \geqslant \mathbf{0}$	Not applicable
Gradients of the objective function at the optimum as a function of the dual variables	$\mathbf{A}^\mathsf{T}\mathbf{u}^* + \nabla f(\mathbf{x}^*) = 0$ $\mathbf{u}^* \geqslant \mathbf{0}$	$\mathbf{A}^\mathsf{T}\mathbf{u}^* + \nabla f(\mathbf{x}^*) = 0$
Sensitivity of the primal variables to small changes in the constraints	Applicable for active constraints only. See next column.	$(\mathrm{d}\mathbf{x}^*/\mathrm{d}\mathbf{b}) =$ $(\nabla^2 f(\mathbf{x}^*))^{-1}\mathbf{A}^\mathsf{T}(\mathbf{A}(\nabla^2 f(\mathbf{x}^*))^{-1}\mathbf{A}^\mathsf{T})^{-1}$
Sensitivity of the dual variables to small changes in the constraints	Applicable for active constraints only. See next column.	$(\mathrm{d}\mathbf{u}^*/\mathrm{d}\mathbf{b}) = -(\mathbf{A}(\nabla^2 f(\mathbf{x}^*))^{-1}\mathbf{A}^\mathsf{T})^{-1}$

Table 3.5 *Table of vector variables*

Variable	Description	Example units
v	Vector of link flows, with generally one element per link, but sometimes more when link flows are disaggregated.	Vehicles per second
c	Vector of link costs, with one element per link.	Dollars per vehicle
s	Vector of link capacities, sometime referred to as link service rates.	Vehicles per second
d	Vector of link delays, with one element per link.	Dollars per vehicle
q	Vector of link queues, with one element per link.	Vehicles
h	Vector of path flows, with generally one element per path, but sometimes more when path flows are disaggregated.	Trips per second
g	Vector of path costs, with one element per path.	Dollars per trip
t	Vector of origin-destination flows, referred to as the trip table. Its elements are referred to as trip rates.	Trips per second
λ	Vector of traffic signal settings	Seconds
u, w	Vectors of dual variables	Various
z	Vector of expected minimum origin–destination costs	Dollars per trip
0, 1, 0.00001	Vectors all of whose elements are 0, 1 and 0.00001 respectively	Real numbers

Table 3.6 *Table of matrix variables*

Variable	Description	Units
A	Link-path incidence matrix, with elements a_{ij} equal to 1 if link i lies on path j, and 0 otherwise.	Binary
B	Origin–destination–path incidence matrix, with elements b_{ij} equal to 1 if origin–destination pair i lies on path j, and 0 otherwise.	Binary
P	Matrix of link choice proportions, with elements p_{ij} equal to the proportion of trips (or probability of a trip) between origin–destination pair j using link i.	Proportion or probability
Φ	Origin–path incidence matrix, with elements ϕ_{ij} equal to 1 if origin i is associated with path j and 0 otherwise.	Binary
Γ	Destination–path incidence matrix, with elements γ_{ij} equal to 1 if destination i is associated with path j and 0 otherwise.	Binary
W	Matrix of exponential terms w_{ij} equal to $\exp(-\alpha c_{ij})$ if there is a link from node i to node j and 0 otherwise. The term c_{ij} represents the cost of traversing the link from node i to node j and α is the dispersion parameter.	Real
N	$N = W + W^2 + W^3 + \ldots = (I - W)^{-1} - I$, provided the sequence is convergent. N has elements v_{ij}.	Real

3.13 *NOTATION*

Tables 3.5 and 3.6 give the default definitions of variables used in this book. Where variables have other or modified definitions, these will be given at the relevant locations in the text. Table 3.5 gives a summary of the vectors most commonly encountered in subsequent chapters. Table 3.6 gives a summary of matrices most commonly encountered in subsequent chapters.

Cost Functions

4.1   INTRODUCTION

Subsequent chapters assume that there is some form of relationship between link flow and link cost, referred to as the *link cost function*. This chapter explores the nature of this relationship. A distinction is made between the costs that arise *between* junctions, referred to here as *link costs*, and costs that arise *at* junctions, referred to here as *junction costs*. In subsequent chapters, the junction costs are projected on to the links to simplify the analysis. This, however, generally leads to *non-separable* link cost functions, as the cost of one link will be correlated with the costs of other links through traffic interactions at the junction. Road traffic networks will be focused on, although occasional reference will also be made to public transport and freight networks.

In practice, there are a number of factors that affect the propensity of a trip-maker to prefer one option over another. The most important factors are travel time, distance, direct cost (fare, toll, fuel consumption, parking charges, etc.), reliability and comfort. Some of these factors, such as travel time, reliability and comfort, are flow related while others are not. Transportation planners and traffic engineers often combine these factors to form a *generalised cost*. This usually has the following linear form:

$$g_j = a_{0j} + a_{1j} \text{ time}_j + a_{2j} \text{ fare}_j + \ldots$$

where $a_{0j}, a_{1j}, a_{2j}, \ldots$ are parameters specific to each option j. In general, the dominant flow-related component of generalised cost is travel time.

4.2   TRAFFIC REPRESENTATION

Link travel times will depend on the dimensions of the road (mainly the number of lanes in each direction), the demand for the link, the way vehicles interact with each other on the link, and any attendant circumstances (like speed limits, parked vehicles, vehicle composition, pedestrian activity near the road, traffic calming measures, etc.) which may influence the speed at which vehicles would like to drive (referred to as the *free flow speed*).

The relationship between link flow and link cost may be established in a number of ways. The best approach is likely to be direct observation, but practical difficulties may arise in making observations. The principal difficulty is obtaining sufficient variation in traffic flow levels, as controlled experiments are generally not feasible. Therefore, to complement direct observations, theories of traffic flow are used. There are a number of different ways of representing and subsequently analysing traffic. Here two ways will be distinguished; the *microscopic representation* and the *macroscopic representation*.

In the microscopic representation, traffic is represented as individual vehicles following simple rules, for example, travelling at a fixed speed, or queuing, or following the vehicle in front, or arriving randomly at a particular location. The microscopic representation is probably best suited to the description of traffic at or in the vicinity of junctions rather than traffic in motion on links.

In the macroscopic representation, traffic is represented as a compressible fluid leading to the application of hydrodynamic theory. Traffic is represented by three variables only, namely flow, speed and density, where flow is equal to speed times density. A relationship between flow and density, referred to as the fundamental diagram, is hypothesised.

4.3 LINK COSTS

4.3.1 Microscopic traffic simulation

A form of computer simulation, referred to as microscopic traffic simulation, is frequently used to study the effect of various factors on traffic parameters, like link travel times. The trajectories of individual vehicles in space and time are generated by an equation that relates the acceleration of a vehicle to the difference in speed between it and the vehicle in front and to the physical headway. A number of microscopic traffic simulators are available, such as VISSIM (Fellendorf, 1994). While microscopic traffic simulation may be used on-line to achieve a functional relationship between link costs and link flows (input link flows, output link costs), this approach is computationally too demanding to be practical. There is therefore an interest in representations of traffic that allow an explicit relationship between flow and cost to be established.

4.3.2 Hydrodynamic traffic theory

One useful representation of traffic, referred to as *hydrodynamic traffic flow theory*, treats traffic as a compressible fluid and describes its properties in terms of *speed, flow* and *density*. This approach goes back to the pioneering work of Lighthill and Whitham (1955) and is set out in detail in a number of text books, in particular Leutzbach (1988).

The basis of the theory is a relationship between traffic flow and traffic density, known as the fundamental diagram (see Fig. 4.1). Flow is zero when density is zero or when density is at a maximum, k_j, referred to as the *queue density*. Note that for any flow, other than maximum flow s, there are two densities. The lower density refers to uncongested flow while the higher density refers to congested flow. In uncongested conditions, flow is generally found to be stable, while in congested conditions, instability prevails. Consider the two densities A and B shown in Fig. 4.1. Since the line connecting A to B is horizontal, the two densities are associated with the same level of flow.

Note that dimensional consistency requires that *flow* (veh/h/lane) be equal to *speed* (km/h) times *density* (veh/km/lane). Therefore the slope of any chord connecting the origin to any point on the fundamental diagram gives the speed corresponding to the flow and density represented by that point. Thus the speed corresponding to density A

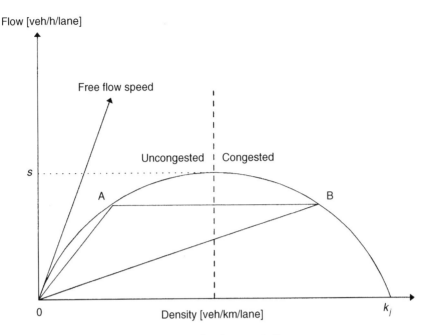

Figure 4.1 *The fundamental diagram*

in Fig. 4.1 is equal to the slope of the line connecting 0 to A, and likewise the speed corresponding to density B is equal to the slope of the line connecting 0 to B.

It is possible to replot a fundamental diagram in terms of a relationship between speed and flow, referred to as a *speed–flow relationship* (see Fig. 4.2). Speed varies between zero and a *free flow speed*. For any given flow other than the maximum, there are two speeds, a speed for uncongested flow and a speed for congested flow. The speed for congested flow is unstable.

Alternatively, a relationship between speed and density may be plotted (see Fig. 4.3). This relationship has been found to be nearly linear in practice, which is useful because linear functions are particularly easy to fit to measurements. The speed–flow relationship may be reinterpreted as a relationship between link travel time and link flow by noting that link travel time is given by link length divided by speed. As time is a major component of travel cost, this relationship provides a basis for a link cost function.

Substantial effort internationally has gone into empirically fitting the fundamental diagram to a variety of situations. Table 4.1 gives a range of empirically verified functions and their proposers.

One feature of the hydrodynamic theory of traffic, explored by Lighthill and Whitham (1955), is the prediction of shock waves. A *shock wave* corresponds to the boundary between traffic of two different densities, like the back or front of a queue, and constitutes a discontinuity of the density surface in the space and time domains. The speed with which a shock wave moves is given by the slope of the cord connecting the points on the fundamental diagram corresponding to the two densities. Figure 4.4 shows a shock wave with speed $(E - F)/(B - A)$.

Strict compliance with the fundamental diagram implicitly supposes that traffic has no momentum, which in turn implies that drivers react instantaneously to any change in

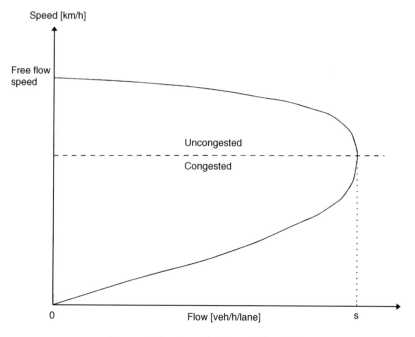

Figure 4.2   *Speed–flow relationship*

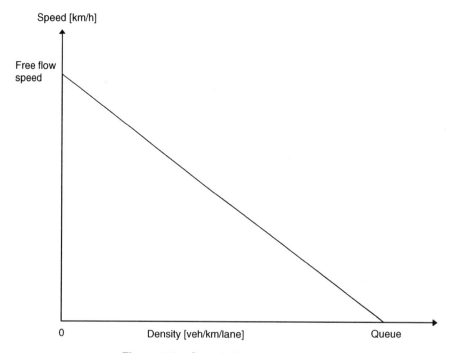

Figure 4.3   *Speed–density relationship*

Table 4.1 *Examples of speed–density functions*

Author	Function (speed in km/h, density in veh/km/lane)
Cremer & Papageorgiou (1981)	speed $= 123(1 - (\text{density}/200)^{1.4})^4$
Papageorgiou *et al.* (1989)	speed $= 90\exp(-0.5(\text{density}/120)^2)$
Payne (1979)	speed $= \min[88.5, 88.5(1.94 - 6(\text{density}/143)$ $+8(\text{density}/143)^2 - 3.93(\text{density}/143)^3]$

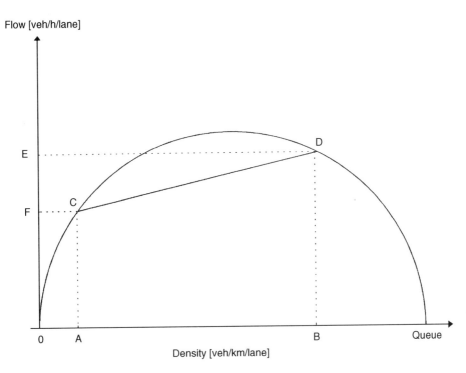

Figure 4.4 *Shock wave with speed (E − F)/(B − A)*

circumstances (for example, to any change in density). Following Payne (1979), it is often assumed that drivers anticipate density changes but react after a small delay, so that actual speed follows the desired speed given by the fundamental diagram, in the following way:

$$w(x, t + \tau) = W(k(x + \delta, t))$$

where τ is the reaction time, δ is the anticipation distance, $w(x, t)$ is the actual speed at point x and time t, and $W(k(x, t))$ is the desired speed expressed as a function of the traffic density at point x and time t. The desired speed is assumed given by the fundamental diagram.

Taylor series expansions of both sides of the above equation lead to the following:

$$w(x, t) + \tau(dw(x, t)/dt) = W(k(x, t)) + \delta(dW(k(x, t))/dk(x, t))(dk(x, t)/dx)$$

To implement the above relationship, the roadway is generally divided into sections of length Δ and time is divided into small steps (of say 1 s). By a finite difference approximation

$$dw(x, t)/dt \approx w_i(t + 1) - w_i(t)$$

Similarly

$$dk(x, t)/dx \approx (k_{i+1}(t) - k_i(t))/\Delta$$

But the rate of change of desired speed with respect to density is more or less constant (see Fig. 4.3) for each section, so

$$dW(k_i(t))/dk_i(t) = u_i$$

where u_i is a section-specific constant. Putting this together yields the following finite difference equation for actual speed in section i, since

$$w_i(t) + \tau(w_i(t + 1) - w_i(t)) = W(k_i(t)) + \delta u_i(k_{i+1}(t) - k_i(t))/\Delta$$

becomes

$$w_i(t + 1) = w_i(t) + (W(k_i(t)) - w_i(t))/\tau + \delta u_i(k_{i+1}(t) - k_i(t))/(\tau\Delta)$$

The second term on the right-hand side is referred to as the *relaxation term* while the third term is referred to as the *anticipation term*. A further term is often added to smooth out speed variations between adjacent road sections (see Papageorgiou *et al.*, 1989).

Input–output relationships may be used to obtain the density for the next period. Thus, in the absence of any external inflows or outflows, the density in section i in period $t + 1$ depends on the inflow from section $i - 1$ in period t and outflow to section $i + 1$ during period t, namely

$$k_i(t + 1) = k_i(t) + k_{i-1}(t)w_{i-1}(t)/\Delta - k_i(t)w_i(t)/\Delta$$

Using these recursive speed and density relationships, a spreadsheet showing how speed and density evolves over space and time may be constructed.

There are a number of minor conceptual difficulties associated with this approach, for example negative speeds are not ruled out. Also the possibility of differences in density (shock waves) propagating forward in space at a speed greater than that of the traffic itself is not ruled out. The latter possibility would imply that drivers respond to conditions behind them, which in general they do not. These issues are dealt with in Daganzo (1995) and Castillo *et al.* (1994).

4.4 JUNCTION COSTS

4.4.1 Queuing theory

The prediction of queue lengths and vehicular delays is important in many aspects of traffic engineering (see Kimber and Hollis, 1979). In the economic appraisal of junction improvement schemes, delay savings form a benefit, and therefore influence whether or not capital expenditure is justified. In predicting the distribution of traffic within urban road networks (a central theme of this book), junction delay is the main factor affecting the route choice of drivers. Hence, junction costs are here equated with junction delays.

Steady state queuing theory is widely used, but predicts infinite queues and delays when the demand reaches the capacity available. In reality, when demand is close to capacity, or when capacity is exceeded for short periods, the queue growth lags behind the expectation of steady state theory and variations over time of demand and capacity cannot be ignored. *Deterministic queuing theory*, in which delay is obtained as the integral of demand minus capacity, can sometimes be used when demand and capacity vary in time. However, this treatment ignores the statistical nature of traffic arrivals and departures, and leads to serious under-estimates of delay unless the capacity is exceeded by a considerable margin. When demand only just reaches capacity zero delays are predicted.

This section deals only with queues of vehicles and corresponding delays, although with some imagination the methods may be extended from queues of vehicles at stop lines to consider the other kinds of queues encountered in transportation networks, for example queues of passengers waiting at a bus stop. Other types of delay, for example geometric delay at a roundabout or a priority junction (that delay suffered in the absence of queues because of the need for vehicles to slow down, negotiate the junction, and accelerate back to normal running speed), are not considered here.

Traffic at a junction may be divided into a number of interacting *streams*, each with a defined capacity and demand. The capacity depends on the space or time available to the stream, and on the type of control. It is influenced by both the road geometry and by the flows in conflicting streams. While capacity is essentially determined by microscopic vehicle interactions, of the kind taken into account in microscopic traffic simulation, it is possible to formulate empirical macroscopic relationships of the form

$$s_i = s_i^0 - \sum_k a_{ik} v_k - \sum_k \sum_l b_{ikl} v_k v_l - \ldots$$

where s_i^0 is the maximum service rate, and a_{ik} and b_{ikl} are parameters (see Kimber and Daly, 1986). Usually a linear model (where b_{ikl} and higher order coefficients are set to zero) is sufficient. At traffic signal junctions, the capacity is determined by the saturation flow during green and the signal timings (see Webster and Cobbe, 1966).

The amount of queuing in a stream depends on the ratio of demand to capacity, referred to as the *traffic intensity*. If this ratio is very much less than unity, queuing is rare, but otherwise it occurs to a greater or lesser extent. Whenever there is queuing, vehicles are delayed.

4.4.2 Steady state theory

Suppose that the demand in a stream is v and that the capacity available is s, both in units of flow (say, vehicles per hour), and that these quantities remain roughly stable over time. If both the arrivals at and the departures from the *stop line* (referred to also as the giveway line or the stop bar) are random, an M/M/1 queuing model is generated. Consider a small interval of time during which the probability of more than one arrival or departure is insignificantly small. Suppose furthermore that there is no serial correlation, namely the actual arrival or departure of a vehicle in one interval does not influence the probability of a vehicle arrival or departure in the next interval. Let α be the probability of an arrival and β be the probability of a departure during this interval. The traffic intensity is then given by

$$\rho = v/s = \alpha/\beta$$

As the system is in a steady state equilibrium, the probability of entering a particular state (in this case, a particular queue size) is equal to the probability of leaving it. Examples of state entering and leaving probabilities are given in Table 4.2.

Since at equilibrium the probability of entering a state should be equal to the probability of leaving it

$$p_1 = p_0(\alpha/\beta)$$

Furthermore after substitution and cancelling

$$p_2\beta = p_1\alpha$$

so

$$p_2 = p_1(\alpha/\beta) = p_0(\alpha/\beta)^2$$

Similarly after substitution and cancelling

$$p_3\beta = p_2\alpha$$

so

$$p_3 = p_2(\alpha/\beta) = p_1(\alpha/\beta)^2 = p_0(\alpha/\beta)^3$$

Note that $\alpha/\beta = \rho$ is the traffic intensity. Thus in general

$$p_n = p_{n-1}\rho = p_0\rho^n$$

As probabilities sum to unity

$$p_0 \sum_{n=0 \ \dots \ N} \rho^n = 1$$

and so

$$p_0 = 1 - \rho$$

Finally by substitution

$$p_n = \rho^n(1 - \rho)$$

The queue length fluctuates in time according to the probability distribution defined above, so p_n represents the proportion of time for which there are n waiting vehicles.

Table 4.2   *State entering and leaving probabilities*

Queue size (state)	Probability of entering state	Probability of leaving state
0	$p_1\beta$	$p_0\alpha$
1	$p_0\alpha + p_2\beta$	$p_1(\alpha + \beta)$
2	$p_1\alpha + p_3\beta$	$p_2(\alpha + \beta)$
etc.		

The average queue length L evaluated over a long period of time can be obtained from

$$L = \sum_{n=0 \text{ to } N} n\,p_n = \rho/(1-\rho)$$

where N is the maximum number of waiting vehicles that can be accommodated, assumed to be a very large number. L includes the vehicle currently being *serviced* (namely the vehicle crossing the stop line); when this vehicle is excluded the average queue length L' is

$$L' = \sum_{n=1-N} (n-1)p_n = L - 1 + p_0 = (2\rho - 1)/(1-\rho) + (1-\rho) = \rho^2/(1-\rho)$$

These expressions are appropriate for the steady state M/M/1 queue with random arrivals and random departures. When the departure process deviates from these assumptions, the Pollaczek–Khintchine formula for the steady state M/G/1 queue

$$L = \rho + C\rho^2/(1-\rho)$$

may be used, where

$$C = 0.5(1 + 1/K)$$

and

$$K = (\text{mean of service time/standard deviation of service time})^2$$

It should be noted that the queue length referred to here does not correspond to the physical queue length, as it assumes that vehicles have no physical size. This is therefore lower than the number of vehicles in the physical queue, because if there were no queue and free flow conditions, there would be vehicles in the space occupied by the physical queue. The physical queue may be estimated as follows:

$$L^{ph} = L/(1 - k/k_j)$$

where L^{ph} is the size of the physical queue, k is the actual density and k_j is the queue density (the average space occupied by a vehicle in a queue is generally taken to be about 6 m).

4.4.3 Time-dependent methods

As the intensity ratio approaches 1, both L and L' tend to infinity. This is of course not what happens during temporary periods of overloading (often referred to as *over-saturation*). Kimber and Hollis (1979) consider how the queue size distribution would evolve over time given an initial queue and a particular intensity ratio. As this produces very mathematically complex expressions, they propose the *coordinate transformation method*.

Let $L(0)$ be the initial queue at time $t = 0$. Then according to deterministic queuing theory,

$$L_d(t) = \text{Max}\{0, (v-s)t + L(0)\} = \text{Max}\{0, (\rho - 1)st + L(0)\}$$

where v is the arrival rate, s the service rate, t the duration of the period and ρ the intensity. In other words, the queue at the end of the period is equal to the queue at the start of the period plus the arrivals minus the departures, or zero, which ever is greater.

The steady state result for queue length, including the vehicle in service, is

$$L = \rho + C\rho^2/(1 - \rho)$$

where C is a constant depending on the arrival and service patterns. For regular arrivals and service, $C = 0$, while for random arrivals and service, $C = 1$. For random arrivals and uniform departures, $C = 0.5$.

The co-ordinate transformation technique is to transform this steady state result so that instead of the queue tending to infinity as ρ approaches 1, it approaches the deterministic value. For any queue size, the steady state intensity ρ_s is transformed to ρ_t so that

$$1 - \rho_s = \rho_d - \rho_t$$

where ρ_d is the intensity corresponding to the queue size in the deterministic case. Hence

$$\rho_d = (L(t) - L(0))/st + 1$$

Thus

$$\rho_s = \rho_t - (L(t) - L(0))/st$$

When $C = 1$, this provides the following transformed relationship

$$L(t) = \rho_s/(1 - \rho_s) = (\rho_t - (L(t) - L(0))/st)/(1 - \rho_t + (L(t) - L(0))/st)$$

This generates a quadratic equation in $L(t)$. Taking the positive route only, the required result is

$$L(t) = 0.5((A^2 + B)^{0.5} - A)$$

where

$$A = (1 - \rho_t)st + 1 - L(0)$$

$$B = 4(L(0) + \rho_t st)$$

The approach is illustrated in Fig. 4.5. The transformation is such that for any queue size, the horizontal distance between the steady state curve and 1, namely AB, is equal to the horizontal distance between the transformed curve and the deterministic queue growth line, namely CD.

Kimber and Hollis (1979) produce the following more general expressions for the case where C is not equal to 1.

$$A = ((1 - \rho)(st)^2 + (1 - L(0))st - 2(1 - C)(L(0) + \rho st))/(st + (1 - C))$$

$$B = (4(L(0) + \rho st)(st - (1 - C)(L(0) + \rho st)))/(st + (1 - C))$$

In the case of *traffic assignment* (a subject at the heart of this book), it is important to have a measure of delay *per arriving vehicle* over an interval t.

The deterministic result is

$$d_d(t) = (L(0) + 1 + 0.5(\rho - 1)st)/s$$

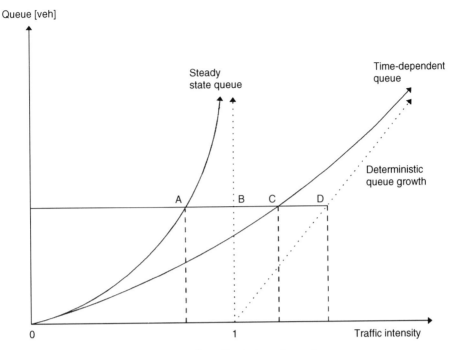

Figure 4.5 *Coordinate transformation*

which is the time it takes the average queue length *plus the arrival* to discharge. The steady state result is

$$d_s = (1 + C\rho^2/(1 - \rho))/s$$

which is also the time it takes the average queue *including the arrival* to discharge. When the vehicle at the stop line is included, the transformed time-dependent result is

$$d(\rho) = 0.5((J^2 + K)^{0.5} - J)$$

where

$$J = 0.5t(1 - \rho) - (L(0) - C + 2)/s$$

$$K = 4(0.5t(1 - \rho) + 0.5\rho tC - (L(0) + 1)(1 - C)/s)/s$$

When the vehicle at the stop line is excluded

$$d'(\rho) = 0.5((P^2 + Q)^{0.5} - P)$$

where

$$P = 0.5t(1 - \rho) - (L(0) - C)/s$$

$$Q = 2Ct((\rho + 2L(0))/st)/s$$

Note that $d(\rho) = d'(\rho) + (1/s)$. This is because on average each vehicle is delayed by at least one service time.

The coordinate transformation expressions just presented are heuristics. In the limiting cases (high or low ρ or t) the expressions are correct, and for intermediate cases, their shapes are reasonable. Kimber and Hollis (1979) have noted a tendency to over-estimate delay. It should be emphasised that the expressions yield *average* queue lengths, and that in reality there will be considerable variation about these average values. This is important in part because of their network effects (for example, large queues may block upstream junctions in urban road networks) and in part because of the effects of risk aversity on decision-making behaviour (for example, on choice of path).

4.5 PRIORITY JUNCTIONS

Conflicts between streams of traffic within the junction reduce the capacities of those streams without priority (it is conventionally assumed that streams with priority experience no delay at the junction). Because the priority streams determine the capacities of the non-priority streams, the cost functions are not separable. The capacity of a non-priority stream may be calculated using linear macroscopic relationships of the form

$$s_i = s_i{}^0 - \sum_{k \in K} a_{ik} v_k$$

where K is the set of conflicting priority streams and $s_i{}^0$ is the capacity of stream i in the absence of any conflicting flows.

Under steady state conditions, the Pollaczek–Khintchine formula for the steady state M/G/1 queue (see Section 4.4.2) may be used with C set at around one. The steady state delay is L/s.

As already noted, the steady state expression does not give a good estimate of delay during periods of transient over-loading. In these circumstances, the co-ordinate transformation method of Kimber and Hollis (1979) set out in Section 4.4.3 may be applied to estimate the delay per arriving vehicle.

Consider a non-priority link with an exit capacity of 15 vehicles per minute. Table 4.3 gives the estimates of the queues and the delays using the Kimber and Hollis (1979) expression for a sequence of one minute intervals, taking C equal to 1. The saturation flow is assumed to be 15 vehicles per minute. The initial queue in the first interval was assumed to be 0, and then L for the preceding interval thereafter. The delay includes the vehicle at the stop line.

4.6 SIGNAL CONTROLLED JUNCTIONS

Consider a link controlled at its exit by traffic signals. For simplicity, the traffic signals are assumed to have just two states; *effective green* when traffic is serviced, and *effective*

Table 4.3 *Queue and delay estimates for a sequence of traffic intensities*

Interval	1	2	3	4	5	6	7	8	9	10
ρ	0.1	0.4	0.7	1.1	1.2	1.0	0.8	0.6	0.4	0.2
Queue	0.103	0.577	1.567	4.508	5.848	4.093	2.483	1.372	0.690	0.278
Delay	0.073	0.102	0.162	0.324	0.530	0.521	0.338	0.199	0.125	0.089

red when no traffic is serviced. Let:

$$r = \text{the red period};$$

$$g = \text{the green period; and}$$

$$c = r + g = \text{the cycle time.}$$

The uniform component of total delay (the component of delay that would arise with a fixed s and v) will then be

$$D = 0.5r^2v/(1 - \rho)$$

provided the queue clears every green period. This is the area of the triangle *ABC* shown in Fig. 4.6. Note the ρ is now the traffic intensity *during the green period* ($\rho \leqslant g/c$). When total delay is divided by the total arrivals over the cycle, the average delay per vehicle as a function of the intensity will be

$$d(\rho) = 0.5r^2/c(1 - \rho)$$

In practice, arrivals are unlikely to be uniform. The randomness of arrivals leads to additional delay, referred to as the random component of delay. This is represented by the area *ABCGFED* in Fig. 4.6. Of critical importance is the expected residual queue at the end of the green period, namely *AD* and *FG* in Fig. 4.6. Denoting the expected residual queue by L, the random component of total delay is approximately equal to Lc. Webster (1958) has therefore proposed adding an additional term to allow for random effects, leading to the following two-term delay formula:

$$d(\rho) = 0.5r^2/c(1 - \rho) + L/v$$

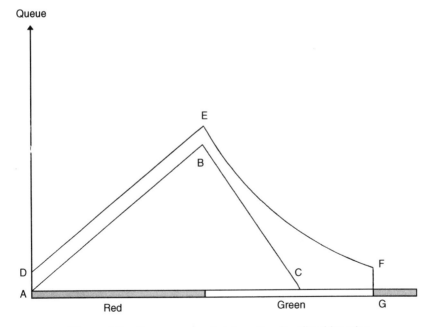

Figure 4.6 *Components of delay at a signalised junction*

where the first term accounts for the uniform component of delay and the second term for the random component of delay. As a first approximation, Webster (1958) suggests that

$$L = x/(1 - x)$$

where in this case x is the *degree of saturation*, defined as follows:

$$x = cv/sg = \rho c/g$$

Simulation experiments led to the addition of a third corrective term.

As indicated earlier, steady state expressions cannot be applied to periods of transient overloading. In these circumstances (or indeed more generally), the Kimber and Hollis (1979) expressions may be used. However, the situation is considerably more complex for signal controlled junctions, and a number of cases have to be considered. The only case considered here is that where no vehicle waits more than one cycle to clear the junction.

The cycle is divided into an effective green period followed by an effective red period. The delay per arriving vehicle over the effective green period is estimated as follows:

$$d_g(\rho) = 0.5((J^2 + K)^{0.5} - J)$$

where

$$J = 0.5g(1 - \rho) - (L(0) - C + 2)/s$$
$$K = 4(0.5g(1 - \rho) + 0.5\rho g C - (L(0) + 1)(1 - C)/s)/s$$

Kimber and Daly (1986) recommend a value for C of 0.5, the traditional value for random arrivals and regular departures. Note that the calculation of the delay per arriving vehicle is based on the time it would take to discharge the average queue at the saturation flow rate.

The residual queue at the end of the green period $L(g)$ may be estimated as follows:

$$L(g) = 0.5((A^2 + B)^{0.5} - A)$$

where

$$A = ((1 - \rho)(sg)^2 + (1 - L(0))sg - 2(1 - C)(L(0) + \rho sg))/(sg + (1 - C))$$
$$B = (4(L(0) + \rho sg)(sg - (1 - C)(L(0) + \rho sg)))/(sg + (1 - C))$$

Over the effective red period, the queue is simply accumulated, so the queue at the end of red is

$$L(c) = L(g) + \rho sr$$

namely the queue at the start of red plus the vehicles predicted to arrive during red. This becomes the queue at the start of the next cycle. The delay per vehicle arriving during effective red is estimated using the following deterministic expression:

$$d_r(\rho) = L(g)/s + 0.5(1 + \rho)r$$

provided that $L(c)/s$ is less than g, namely that all the vehicles arriving in one cycle can leave by the end of the next. The calculation of the delay during red is illustrated in Fig. 4.7. A vehicle arriving at the start of red must wait $L(g)/s + r$, namely the time

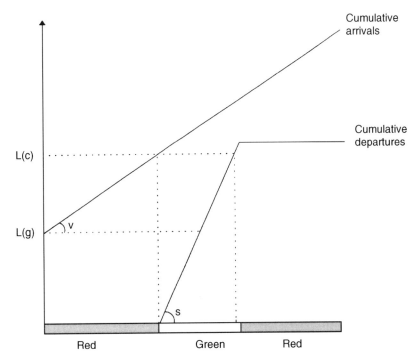

Figure 4.7   *The delay to arriving vehicles during red if junction under-saturated*

it takes to service the queue already there plus the red time. A vehicle arriving at the end of red must wait $L(c)/s = L(g)/s + \rho r$, namely the time it takes to service the queue already there plus the arrivals during red. The average delay per vehicle for a vehicle arriving during red is given as the average of these two, namely $L(g) + 0.5r(1 + \rho)$.

When $L(c)/s$ is greater than g, some (or possibly all) vehicles arriving during red will not go through on the next green. This will significantly increase the average delay per vehicle for a vehicle arriving during red, as shown in Fig. 4.8 by the horizontal gap between the cumulative arrivals and cumulative departures curves.

The average delay per vehicle arriving over the cycle is the average delay for a vehicle arriving during green times the probability of arriving during green plus the average delay for a vehicle arriving on red times the probability of arriving on red, namely

$$d(\rho) = (gd_g(\rho) + rd_r(\rho))/c$$

The residual queue $L(c)$ then becomes the initial queue $L(0)$ for the next cycle.

Consider a signal controlled stream with a cycle time of 1 minute, a saturation flow of 30 vehicles per minute and a 30 second average green time. Table 4.4 shows the evolution

Table 4.4   *Queue and delay estimates for a sequence of degrees of saturation*

Interval	1	2	3	4	5	6	7	8	9	10
ρ'	0.1	0.4	0.7	1.1	1.2	1.0	0.8	0.6	0.4	0.2
$L(c)$	0.798	3.255	5.816	9.208	10.043	8.384	6.762	5.118	3.458	1.786
Delay	0.165	0.195	0.238	0.312	0.375	0.362	0.313	0.269	0.233	0.215

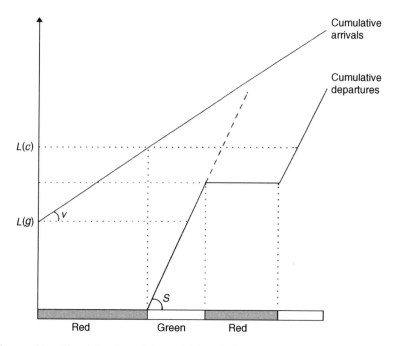

Figure 4.8   *The delay to arriving vehicles during red if junction over-saturated*

of queues *at the end of red*, namely $L(c)$, and the delays per arriving vehicle as predicted by the above expression. The initial queue in the first interval was assumed to be 0, and then $L(c)$ for the preceding interval thereafter. The delay includes the vehicle at the stop line.

4.7 DISCUSSION

This chapter has reviewed approaches to calculating link costs. A distinction has been drawn between costs that arise due to *flow* between junctions and costs that arise due to *queuing* at junctions. During periods of transient overload that characterise urban congestion, time-dependent expressions are required. The Kimber and Hollis (1979) co-ordinate transformation method for mapping steady state queuing expressions on to time-dependent axes has been presented. The Kimber and Hollis approach fits well with the time-dependent formulations in later chapters, where steady state time slices linked by equilibrium queues are considered.

5

Deterministic User Equilibrium Assignment

5.1 INTRODUCTION

The process whereby demand in the form of a trip table is loaded on to the network is referred to as *traffic assignment*. The process of loading should reflect rational decision-making behaviour by trip-makers. Rationality comes in different forms, depending on whose interests are being pursued. Altruistic rationality would require trip-makers to behave in a way that maximises the benefit to the community, leading to a *system optimum*. Individualistic rationality, on the other hand, would require trip-makers to pursue their own interests individually (or selfishly) without regard to the interests of others, leading to a *user equilibrium*. For example, it would be individualistically rational for each trip-maker to select what he or she perceived to be the least cost path. A system optimal assignment may be regarded as a *prescriptive* assignment while a user equilibrium assignment may be regarded as a *descriptive* assignment.

The assumption of individualistic rationality in traffic assignment is frequently attributed to Wardrop (1952) and known as Wardrop's first principle. As a behavioural proposition it is plausible. At the user equilibrium, paths are chosen so that no trip-maker *individually* can obtain an advantage by changing path. This assignment is to be distinguished from a *user optimum* assignment, because situations can arise where, if *some* trip-makers are persuaded to act in a way that initially seems to be against their own interests, *all* trip-makers gain (see Heydecker, 1986, for a discussion on the nature of equilibrium). Braess's paradox (Braess, 1968) is an example of such a situation. Braess's example relates to user equilibrium assignment in a five-link network with one origin–destination pair and link cost functions that are separable and monotonically increasing. The paradox arises because the removal of one link and the resulting reassignment of traffic causes the origin–destination cost to fall. A numerical example of Braess's paradox is to be found in Section 2.13.

In this and the subsequent chapter, two forms of traffic assignment are examined. Both presuppose that trip-makers are individualistically rational (in short, selfish) in their decision-making behaviour. The first is *deterministic user equilibrium* assignment, which assumes that trip-makers perceive costs identically, in effect requiring perfect (or at least uniform) foresight and rules out *errors in perception*. The second is *stochastic user equilibrium* assignment, which relaxes the assumption of the identical perceptions of costs, and therefore allows for errors in perception.

Additionally, two forms of link cost function are considered. The first has link cost increasing as a function of link flow without a maximum flow (although there may be a level of flow beyond which the costs would be *unreasonable* for any trip-maker to pay). This is referred to here as an *increasing cost function*. The second has a constant cost and a fixed capacity. At the capacity, queuing occurs on the link and the cost is determined so as to ensure that the demand does not exceed the capacity. This is referred to here as a *right angle cost function*.

Consider the two-link, two-path network considered in Fig. 5.1. The principle of a deterministic user equilibrium assignment is illustrated for increasing cost functions in Fig. 5.2 and for right angle cost functions in Fig. 5.3. The equilibrium division of traffic between the two paths is given by the intersection of the two cost curves. For a division of traffic to the *left* of the equilibrium, link 1 is less costly than link 2, providing an incentive for traffic to switch from link 2 to link 1, which in turn will move the division of traffic to the *right*. For a division of traffic to the *right* of the equilibrium, link 1 is more costly than link 2, providing an incentive for traffic to switch from link 1 to link 2, which in turn will move the division of traffic to the *left*. The equilibrium is therefore *stable*, in the sense that any deviation from the equilibrium generates incentives that act to restore the equilibrium.

The response surface is shown in Fig. 5.4. This has a step at the point where the two paths have equal cost. At this point, the choice probability is undefined. This illustrates

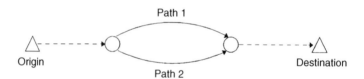

Figure 5.1   *A two-link network*

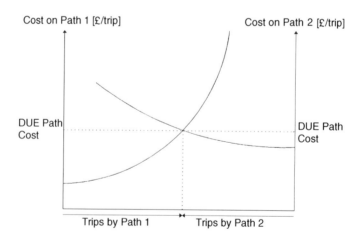

Figure 5.2   *Deterministic user equilibrium for increasing cost functions*

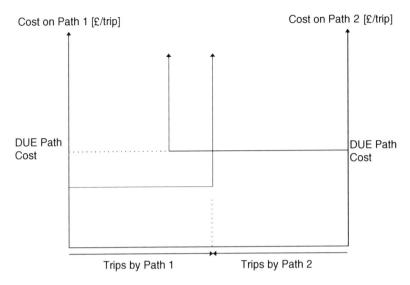

Figure 5.3 *Deterministic user equilibrium for right angle cost functions*

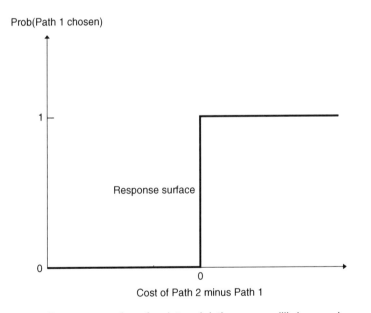

Figure 5.4 *Response surface for deterministic user equilibrium assignment*

one of the difficulties with deterministic user equilibrium assignment, namely that in equilibrium all the used paths between a given origin and a given destination have an equal cost, but when path costs are equal the response surface shows that the share of traffic between the paths may not be defined. Thus in general path flows are not uniquely defined in deterministic user equilibrium.

5.2 *EXISTENCE AND UNIQUENESS*

Consider first the case where link costs are a function of link flows, such that for every feasible link flow vector we can associate a unique link cost vector. This rules out the right angle cost function. Then

$$\mathbf{c} = \mathbf{c}(\mathbf{v})$$

is a differentiable (it is assumed here twice differentiable) function, where $\mathbf{c}$ and $\mathbf{v}$ are vectors of link costs and link flows respectively. The form of this function has been discussed in Chapter 4. Link flows are related to path flows by the link–path incidence matrix as follows:

$$\mathbf{v} = \mathbf{A}\mathbf{h}$$

where $\mathbf{A}$ is the link–path incidence matrix introduced in Chapter 2 and $\mathbf{h}$ is a vector of path flows. Let $\mathbf{g}$ be the vector of path costs. Then since the cost of any path consists of the sum of the costs of the constituent links

$$\mathbf{g}(\mathbf{h}) = \mathbf{A}^{\mathrm{T}}\mathbf{c}(\mathbf{v})$$

Clearly path costs are functions of link flows and therefore of path flows, as follows:

$$\mathbf{g} = \mathbf{A}^{\mathrm{T}}\mathbf{c}(\mathbf{A}\mathbf{h})$$

Let $\mathbf{h}^*$ be the user equilibrium path flows and $\mathbf{g}(\mathbf{h}^*)$ the corresponding path costs. Then under deterministic user equilibrium assignment the following variational inequality holds:

$$(\mathbf{h} - \mathbf{h}^*)^{\mathrm{T}}\mathbf{g}(\mathbf{h}^*) \geqslant 0$$

where $\mathbf{h}$ is any other vector of feasible path flows (see Smith, 1979). The variational inequality simply says that, *given user equilibrium path costs*, any deviations from the user equilibrium path flows cannot reduce total costs. In other words, no trip-maker would be *tempted* to reduce his cost by changing his path, as no alternative path would offer a lower cost. Of course, whether or not any trip-maker could actually reduce his costs is another matter (see Heydecker, 1986).

The variational inequality can be restated in terms of link flows as follows:

$$(\mathbf{h} - \mathbf{h}^*)^{\mathrm{T}}\mathbf{g}(\mathbf{h}^*) = (\mathbf{v} - \mathbf{v}^*)^{\mathrm{T}}\mathbf{A}^{\mathrm{T}}\mathbf{g}(\mathbf{h}^*) = (\mathbf{v} - \mathbf{v}^*)^{\mathrm{T}}\mathbf{c}(\mathbf{v}^*) \geqslant 0$$

where $\mathbf{v}^*$ is the vector of deterministic user equilibrium link flows, $\mathbf{c}(\mathbf{v}^*)$ the corresponding vector of link costs and $\mathbf{v}$ any other vector of feasible link flows.

The constraints take the form of the trip table and non-negativity conditions on the paths flows, namely

$$\mathbf{t} = \mathbf{B}\mathbf{h}, \qquad \mathbf{h} \geqslant \mathbf{0}$$

where $\mathbf{t}$ is the trip table in vector form and $\mathbf{B}$ is the origin–destination–path incidence matrix introduced in Chapter 2.

The constraints are convex in both path flows and link flows. Concerning the path flows, suppose $\mathbf{h}_1$ and $\mathbf{h}_2$ are two vectors of feasible path flows, then $\mathbf{h}_1\lambda + \mathbf{h}_2(1 - \lambda)$ is also feasible for $1 \geqslant \lambda \geqslant 0$ because

$$\mathbf{B}(\mathbf{h}_1\lambda + \mathbf{h}_2(1 - \lambda)) = \mathbf{t}\lambda + \mathbf{t}(1 - \lambda) = \mathbf{t}$$

Suppose $\mathbf{v}_1$ and $\mathbf{v}_2$ are two vectors of feasible link flows, then there must exist two vectors of feasible path flows, $\mathbf{h}_1$ and $\mathbf{h}_2$, such that

$$\mathbf{v}_1\lambda + \mathbf{v}_2(1 - \lambda) = \mathbf{A}\mathbf{h}_1\lambda + \mathbf{A}\mathbf{h}_2(1 - \lambda) = \mathbf{A}(\mathbf{h}_1\lambda + \mathbf{h}_2(1 - \lambda))$$

Since $\mathbf{h}_1\lambda + \mathbf{h}_2(1 - \lambda)$ is feasible for $1 \geqslant \lambda \geqslant 0$ so is $\mathbf{v}_1\lambda + \mathbf{v}_2(1 - \lambda)$. Note also that $\mathbf{h} \geqslant \mathbf{0}$ implies that $\mathbf{v} \geqslant \mathbf{0}$, because the elements of $\mathbf{A}$ are non-negative.

As stated in Chapter 2, matrix $\mathbf{A}$ has elements a_{ij} equal to 1 if link i lies on path j and 0 otherwise. When link i lies on path j, all the flow on path j lies on link i but in general not all the flow on link i lies on path j. Therefore

$$a_{ij}v_i \geqslant h_j$$

and hence

$$\mathbf{A}^T\mathbf{v} \geqslant \mathbf{h}$$

which after pre-multiplication by matrix $\mathbf{B}$ yields the following convex constraints for link flows:

$$\mathbf{B}\mathbf{A}^T\mathbf{v} \geqslant \mathbf{t}$$

However, these constraints are of limited usefulness as they only partially represent the constraints $\mathbf{v} = \mathbf{A}\mathbf{h}$, $\mathbf{t} = \mathbf{B}\mathbf{h}$ and $\mathbf{h} \geqslant \mathbf{0}$.

The existence of an equilibrium depends on very general conditions (see Smith, 1979). By setting up an equivalent linear programming problem, it is demonstrated here that the conditions required for the existence of a deterministic user equilibrium assignment are that the set of feasible path flows be non-empty.

Taking any vector of link costs $\mathbf{c}$, consider the following linear programming problem:

$$\text{Minimise } \mathbf{h}^T(\mathbf{A}^T\mathbf{c}) \text{ subject to } \mathbf{t} = \mathbf{B}\mathbf{h}, \; \mathbf{h} \geqslant \mathbf{0}$$

If the set of feasible path flows is non-empty, this problem has a solution, say $\mathbf{h}^*$. As linear programing problems are also convex programming problems, the optimal path flows are fully (but not usually uniquely) identified by the following inequality.

$$(\mathbf{h} - \mathbf{h}^*)^T\mathbf{A}^T\mathbf{c} = (\mathbf{v} - \mathbf{v}^*)^T\mathbf{c} \geqslant 0 \qquad \text{for all feasible } \mathbf{v}$$

(see Chapter 3 on optimality conditions for convex programming problems). Setting $\mathbf{c}$ equal to $\mathbf{c}(\mathbf{v}^*)$ ensures the *existence* of a deterministic user equilibrium assignment, provided the set of feasible path flows is non-empty.

Given the existence of a deterministic user equilibrium assignment, the positive definiteness of the Jacobian of the link costs, namely $\nabla\mathbf{c}(\mathbf{v})$, is *necessary* and *sufficient* for uniqueness. Suppose there are two distinct equilibria, $\mathbf{v}^*$ and $\mathbf{v}^{**}$, with $\mathbf{v}^* \neq \mathbf{v}^{**}$. The positive definiteness of the Jacobian of $\mathbf{c}(\mathbf{v})$ ensures that

$$(\mathbf{v}^* - \mathbf{v}^{**})^T(\mathbf{c}(\mathbf{v}^*) - \mathbf{c}(\mathbf{v}^{**})) > 0$$

As both $\mathbf{v}^*$ and $\mathbf{v}^{**}$ are equilibria

$$(\mathbf{v}^{**} - \mathbf{v}^*)^T\mathbf{c}(\mathbf{v}^*) \geqslant 0$$

and

$$(\mathbf{v}^* - \mathbf{v}^{**})^T\mathbf{c}(\mathbf{v}^{**}) \geqslant 0$$

Together, these suggest that

$$(\mathbf{v}^* - \mathbf{v}^{**})^T(\mathbf{c}(\mathbf{v}^*) - \mathbf{c}(\mathbf{v}^{**})) \leqslant 0$$

which contradicts the positive definiteness of the Jacobian of link costs. Hence the solution is *unique* in the link flows (but not necessarily in the path flows).

An *equivalent optimisation problem* is of particular interest from the point of view of finding a deterministic user equilibrium assignment for large transportation networks. If there is an optimisation problem of the following form:

$$\text{Minimise } f(\mathbf{v}) \text{ subject to } \mathbf{v} = \mathbf{Ah}, \mathbf{t} = \mathbf{Bh} \text{ and } \mathbf{h} \geqslant \mathbf{0}$$

then its solution would be the required assignment if $f(\mathbf{v})$ were convex with

$$\nabla f(\mathbf{v}) = \mathbf{c}(\mathbf{v})$$

because then at the optimum

$$(\mathbf{v} - \mathbf{v}^*)^T \mathbf{c}(\mathbf{v}^*) \geqslant 0 \qquad \text{for all feasible } \mathbf{v}$$

since the constraints on $\mathbf{v}$ are, as already noted, convex (see Chapter 3 for the optimality conditions). If the Jacobian is positive definite for all link flows, then the solution is also unique, as shown above (see also Chapter 3).

Note that $\nabla^2 f(\mathbf{v})$, and therefore $\nabla \mathbf{c}(\mathbf{v})$, must be symmetric as it is the Hessian of $f(\mathbf{v})$ and

$$\partial^2 f(\mathbf{v})/\partial v_i \partial v_j = \partial^2 f(\mathbf{v})/\partial v_j \partial v_i \qquad \text{for all links } i \text{ and } j$$

When the link cost functions are separable and monotonically increasing

$$\partial^2 f(\mathbf{v})/\partial v_i \partial v_j = \partial c_i(\mathbf{v})/\partial v_j = \begin{cases} > 0 & \text{if } i = j \\ 0 & \text{if } i \neq j \end{cases}$$

which implies that $\nabla^2 f(\mathbf{v})$ is positive definite and that the assignment is unique in terms of the link flows.

With right angle cost functions, the variational inequality has the form

$$(\mathbf{v} - \mathbf{v}^*)^T \mathbf{c} \geqslant 0$$

and the constraints are

$$\mathbf{v} = \mathbf{Ah}, \qquad \mathbf{t} = \mathbf{Bh}, \qquad \mathbf{h} \geqslant \mathbf{0}, \qquad \mathbf{s} \geqslant \mathbf{v}$$

where $\mathbf{s}$ is the vector of link capacities. As the cost functions are separable, there is an equivalent optimisation problem, namely

$$\text{Minimise } f(\mathbf{v}) \text{ subject to } \mathbf{v} = \mathbf{Ah}, \mathbf{t} = \mathbf{Bh}, \mathbf{h} \geqslant \mathbf{0}, \mathbf{s} \geqslant \mathbf{v}$$

where

$$\nabla f(\mathbf{v}) = \mathbf{c}$$

Note that

$$\nabla^2 f(\mathbf{v}) = \nabla \mathbf{c} = \mathbf{0}$$

which is not positive definite. By an argument similar to that made for increasing cost functions, the existence of a deterministic user equilibrium assignment may be shown for the case of right angle cost functions. However, the Hessian of the equivalent optimisation problem is not positive definite, so the assignment is in general not unique.

In summary, the *existence* of a deterministic user equilibrium assignment is assured if the constraints are convex, which in general they are. The *uniqueness* of deterministic user equilibrium link flows is guaranteed if the Jacobian of the link cost functions with respect to link flows, namely $\nabla c(v)$, is positive definite. The implications of this uniqueness condition are that *the dominant effect on the cost of any link i must be the flow on link i, and this effect must be monotonically increasing.* This does not rule out the possibility of the flow on links other than i affecting the cost on link i, as is generally the case in urban road networks, but this effect should be of secondary significance. In the case of right angle cost functions, the Jacobian of the link cost functions with respect to link flows is zero, implying the existence but possible non-uniqueness of deterministic user equilibrium link flows.

In order to formulate an equivalent optimisation problem, the Jacobian of the link cost functions with respect to link flows should be *symmetric*, in the sense that a small change in the flow on link i should affect the cost on link j in the same way as a small change in flow on link j would affect the cost on link i. In practice, symmetry in the link cost functions seems rather unlikely. The presence of asymmetry does not, however, seem to much affect the performance of algorithms based on equivalent optimisation problems.

Even where there is a unique vector of deterministic user equilibrium link flows, there will in general be many (indeed infinitely many) feasible path flow vectors. This is because in general there are many more paths in a network than links. This fact is overlooked in some software packages which produce estimates of path flows, perhaps in the hope that the deterministic user equilibrium path flow vectors do not differ significantly from one another.

5.3 SOLUTION ALGORITHMS

5.3.1 Increasing cost functions

The variational inequality condition set out in the preceding section does not help much in finding an equilibrium assignment, although it may be applied to confirm when one has been found. When the cost functions are *separable* (namely, when the cost on link i depends only on the flow on link i), it is possible to write the objective function of an equivalent optimisation problem as

$$f(v) = \sum_{i=1 \text{ to } N} \int_{x=0}^{x=v_i} c_i(x)\, dx$$

(where N is the number of links), since the gradient vector $\nabla f(v)$ has elements $c_i(v_i)$ for $i = 1$ to N. If $c_i(v_i)$ is monotonically increasing for all i then $\nabla^2 f(v)$ is positive definite and $f(v)$ is convex. The above objective function was first introduced into the transportation literature by Beckmann *et al.* (1956).

If the link cost functions are monotonic, they may be inverted, since one and only one cost corresponds to any flow. Thus integration by parts yields an alternative objective function

$$f(\mathbf{v}) = \sum_{i=1 \text{ to } N} \int_{x=0}^{x=v_i} c_i(x)\,dx = \mathbf{v}^T\mathbf{c}(\mathbf{v}) - \sum_{i=1-N} \int_{x=c_{min}}^{x=c_i(\mathbf{V})} v_i(x)\,dx = f'(\mathbf{c}(\mathbf{v}))$$

This alternative objective function is rarely considered in the literature.
An efficient approach to solve the following equivalent optimisation problem

Minimise $f(\mathbf{v})$ subject to $\mathbf{v} = \mathbf{Ah}, \mathbf{t} = \mathbf{Bh}, \mathbf{h} \geqslant \mathbf{0}$

is offered by the Frank–Wolfe algorithm (originally proposed by Bruynooghe *et al*., 1968).
Following the determination of an initial set of feasible link flows (Step 1), the algorithm
repeatedly solves a linear programming problem to obtain *auxiliary link flows* (Step 2)
and performs a *line search* for the optimal convex combination of the auxiliary link flows
and the current link flows (Step 3), resulting in an update of the current link flows. The
steps in more detail are:

*Frank—Wolfe algorithm for deterministic user equilibrium
assignment*

Step 1   (initialisation)
         $\mathbf{v}' \leftarrow$ Minimise $LP(\mathbf{v}) = \mathbf{v}^T\nabla f(\mathbf{0})$ subject to $\mathbf{v} = \mathbf{Ah}, \mathbf{t} = \mathbf{Bh}, \mathbf{h} \geqslant \mathbf{0}$

Step 2   (auxiliary link flows)
         $\mathbf{v}^* \leftarrow$ Minimise $LP(\mathbf{v}) = \mathbf{v}^T f(\mathbf{v}')$ subject to $\mathbf{v} = \mathbf{Ah}, \mathbf{t} = \mathbf{Bh}, \mathbf{h} \geqslant \mathbf{0}$

Step 3   (line search)
         $\mathbf{v}' \leftarrow$ Minimise $f(\mathbf{v} = \mathbf{v}'\lambda + \mathbf{v}^*(1 - \lambda))$ subject to $1 \geqslant \lambda \geqslant 0$
         if convergence insufficient then return to Step 2, else stop.

Note that the linear programming problems posed in Steps 1 and 2 are equivalent
to finding all-or-nothing assignments, namely the building of least cost paths and the
loading of the traffic on to these paths. In Step 3, a convex combination of the auxiliary
and current link flows is then sought to minimise the objective function. This is an
optimisation problem in one variable, λ, for which a number of techniques are available,
for example the *bisection method*, if the derivatives of the objective function with respect
to λ are available, or the *Golden Section* search method otherwise. The result of this
optimisation is a new vector of current link flows. If convergence is incomplete after
Step 3, the algorithm returns to Step 2 and seeks a new vector of auxiliary link flows on
the basis of the current link costs, etc.

There are a number of ways of assessing the degree of convergence. One way is
to observe the relative changes in the vector of current link flows between iterations.
Another way, making use of the optimality condition, is to compare $\mathbf{v}^{*T}\mathbf{c}(\mathbf{v}')$ with $\mathbf{v}'^T\mathbf{c}(\mathbf{v}')$.
Convergence is achieved when $\mathbf{v}'$ changes little between iterations and when $\mathbf{v}^{*T}\mathbf{c}(\mathbf{v}')$ is
close to $\mathbf{v}'^T\mathbf{c}(\mathbf{v}')$.

The beauty of the algorithm from the point of view of the transportation planner or
traffic engineer is its relative simplicity and its use of simple and well-understood steps.
The all-or-nothing assignment technique is familiar to many practitioners. The line search
step is also simple to understand. Furthermore, the method is economical in terms of
storage. Although the paths are built in Step 2, it is not necessary to store these as the
link–path incidence matrix is not required in Step 3.

If the line search should prove to be a problem, this may be replaced by the method of successive averages, whereby

$$\mathbf{v}' \leftarrow \mathbf{v}'(1 - 1/n) + \mathbf{v}^*(1/n)$$

with n being the iteration number. The method of successive averages is described in Section 3.10. On the first iteration, $\mathbf{v}' \leftarrow \mathbf{v}^*$; on the second iteration, $\mathbf{v}' \leftarrow \mathbf{v}'(\frac{1}{2}) + \mathbf{v}^*(\frac{1}{2})$; etc. The Frank–Wolfe algorithm with either a line search or the method of successive averages will converge, provided there is a unique solution in terms of link flows (see the preceding section on uniqueness). Where a line search is performed, convergence is assured by the fact that all search directions are descent directions and all steps are descent steps. In the case of the method of successive averages, convergence has been proven by Powell and Sheffi (1982). The speed of convergence may, however, be a problem with either a line search or the method of successive averages.

Various proposals have been made for accelerating the Frank–Wolfe algorithm, focusing mainly on Step 3. Perhaps the most promising among these is the *simplicial decomposition* approach (see Patriksson, 1994). At the expense of retaining all the auxiliary flow vectors generated so far by Step 2, Step 3 becomes

$$\mathbf{v}' \leftarrow \text{Minimise} f \left(\mathbf{v} = \sum_{i=1 \text{ to } n} \mathbf{v}^*{}_i \lambda_i \right) \text{ subject to } \sum_{i=1 \text{ to } n} \lambda_i = 1$$

where, as before, n is the iteration number. The new vector of current link flows is arrived at by seeking the optimal convex combination of the previous vectors of auxiliary link flows. A variety of numerical methods may be used here, for example Newton's method or a quasi-Newton method.

In practice, link cost functions are unlikely to be separable, particularly in urban road networks. The favoured approach for dealing with non-separable cost functions appears to be the Frank–Wolfe algorithm with *diagonalisation*, by which is meant the use of $\mathbf{c}(\mathbf{v}')$ instead of $\nabla f(\mathbf{v}')$ in Step 2 and the consideration of the $c_i(v_i)$ terms only in Step 3, as is required by the Beckmann objective function. In practice, diagonalisation appears to work well (see Meneguzzer, 1995).

As an example of the Frank–Wolfe algorithm in operation, consider the network in Fig. 7.1. Each link has the same cost function, namely

$$\text{cost} = 1 + (\text{flow})^2$$

The origin–destination demands from A to C, A to D, B to C and B to D are assumed to be 10 units of flow, say vehicles per minute, in each case. As a result of the symmetry of the problem, the equilibrium link flows are 20, 20, 5, 5, 5 and 5 for links 1, 2, 3, 4, 5 and 6 respectively. The path with links 3, 2 and 6 as well as the path with links 4, 1 and 5 are not used; also paths with cycles would not be used. Table 5.1 gives the link flows obtained from the Frank–Wolfe algorithm for varying numbers of iterations, when the method of successive averages is used to combine current and auxiliary link flows. Convergence is fairly slow, and after 200 iterations there is still a significant deviation from the equilibrium values.

When the method of successive averages is replaced by a line search (in this case using the bisection method), Table 5.2 indicates that convergence in the early iterations of the algorithm is assisted somewhat.

Table 5.1   *Frank–Wolfe algorithm with method of successive averages*

Iterations	Flows (veh/min)					
	Link 1	Link 2	Link 3	Link 4	Link 5	Link 6
10	21	19	7	8	8	7
20	20.5	19.5	6	6.5	6.5	6
50	20.2	19.8	5.4	5.6	5.6	5.4
100	19.9	20.1	5.3	5.2	5.2	5.3
200	20.05	19.95	5.10	5.15	5.15	5.10

Table 5.2   *Frank–Wolfe algorithm with bisection method*

Iterations	Flows (veh/min)					
	Link 1	Link 2	Link 3	Link 4	Link 5	Link 6
10	20.18	19.81	6.00	6.19	6.19	6.00
20	20.13	19.88	5.75	5.88	5.88	5.75
50	20.07	19.93	5.44	5.51	5.51	5.44
100	20.04	19.96	5.27	5.31	5.31	5.27
200	20.02	19.98	5.16	5.18	5.18	5.16

5.3.2   Right angle cost functions

A special case arises when each link has a constant cost and a fixed capacity. When the demand for a link exceeds its capacity, an extra cost is incurred (referred to as *delay*) which is sufficient to balance the supply (in this case, the capacity) and the demand. Note that when link costs are constant

$$f(\mathbf{v}) = \sum_{i=1 \text{ to } N} \int_{x=0}^{x=v_i} c_i(x)\,\mathrm{d}x = \sum_{i=1 \text{ to } N} c_i v_i = \mathbf{c}^{\mathrm{T}}\mathbf{v}$$

Hence a deterministic user equilibrium assignment may be obtained as a solution to the following linear programming problem:

Minimise $\mathbf{v}^{\mathrm{T}}\mathbf{c}$ subject to $\mathbf{v} = \mathbf{A}\mathbf{h}$, $\mathbf{t} = \mathbf{B}\mathbf{h}$, $\mathbf{h} \geqslant 0$, $\mathbf{s} \geqslant \mathbf{v}$

where $\mathbf{c}$ is a vector of fixed link costs and $\mathbf{s}$ is a vector of link capacities. As before, this may be reformulated in terms of path flows to give

Minimise $\mathbf{h}^{\mathrm{T}}(\mathbf{A}^{\mathrm{T}}\mathbf{c})$ subject to $\mathbf{t} = \mathbf{B}\mathbf{h}$, $\mathbf{h} \geqslant 0$, $\mathbf{s} \geqslant \mathbf{A}\mathbf{h}$

The least cost paths are filled first, followed by the second best paths, etc., until all the trip table has been loaded. No individual trip-maker will be able to save by changing his path, as all lesser cost paths (if any) are already full. In this sense, the result will be a deterministic user equilibrium assignment.

An interesting property of this approach is that the dual variables of the capacity constraints give the equilibrium link delays (see Smith, 1987, and Bell, 1995a). In Chapter 3, it was shown that a dual variable gives the effect on the objective function of a marginal relaxation of the corresponding constraint. If a link capacity constraint is inactive, the corresponding dual variable is zero. If the capacity constraint is active, the relaxation will allow one unit of flow (say one trip) on some more costly path to swap to some less costly path. In order to minimise the objective function, the swap should be between the pair of paths offering the largest cost saving. The value of the dual variable is then equal to this saving. It follows that the dual variable can be interpreted as the delay at the deterministic user equilibrium, since in this case *all used paths are least cost paths, and all paths that are not least cost paths are not used.*

Unfortunately, the Frank–Wolfe algorithm cannot be used here, because link cost is undefined at link capacity. When the paths are defined beforehand, the path-based linear programming problem may be solved using standard procedures. For the link-based linear programming problem, a column generation scheme must be employed.

Consider the path-based linear programming problem

$$\text{Minimise } \mathbf{g}^T\mathbf{h} \text{ subject to } \mathbf{t} = \mathbf{Bh}, \mathbf{s} \geqslant \mathbf{Ah}, \mathbf{h} \geqslant \mathbf{0}$$

The Lagrangian equation is

$$L(\mathbf{h}, \mathbf{z}, \mathbf{u}, \mathbf{k}) = \mathbf{g}^T\mathbf{h} + \mathbf{z}^T(\mathbf{t} - \mathbf{Bh}) - \mathbf{u}^T(\mathbf{s} - \mathbf{Ah}) - \mathbf{k}^T\mathbf{h}$$

where $\mathbf{z}$, $\mathbf{u}$ and $\mathbf{k}$ are vectors of dual variables. At the optimum

$$\nabla_h L(\mathbf{h}, \mathbf{z}, \mathbf{u}, \mathbf{k}) = \mathbf{g} - \mathbf{B}^T\mathbf{z}^* + \mathbf{A}^T\mathbf{u}^* - \mathbf{k}^* = \mathbf{0}$$

$$\nabla_u L(\mathbf{h}, \mathbf{z}, \mathbf{u}, \mathbf{k}) = \mathbf{s} - \mathbf{Ah}^* \geqslant \mathbf{0}$$

$$\mathbf{u}^* \geqslant \mathbf{0}$$

$$\mathbf{u}^{*T}(\mathbf{s} - \mathbf{Ah}^*) = 0$$

$$\nabla_k L(\mathbf{h}, \mathbf{z}, \mathbf{u}, \mathbf{k}) = \mathbf{h} \geqslant \mathbf{0}$$

$$\mathbf{k}^* \geqslant \mathbf{0}$$

$$\mathbf{k}^{*T}\mathbf{h} = 0$$

where the subscript attached to the partial differentiation operator ∇ indicates by which variable the partial differentiation is performed. At the optimum, vector $\mathbf{u}^*$ gives the deterministic user equilibrium link delays, so vector $\mathbf{A}^T\mathbf{u}^*$ contains the deterministic user equilibrium path delays. Vector $\mathbf{z}^*$ gives the cost incurred by the *marginal* (or last) unit of flow between each origin–destination pair. Therefore $\mathbf{B}^T\mathbf{z}^*$ is the vector of deterministic user equilibrium path costs *inclusive of delay* for all used paths. For all paths,

$$\mathbf{A}^T\mathbf{u}^* + \mathbf{g} \geqslant \mathbf{B}^T\mathbf{z}^*$$

For used paths, this is an equality. Consider only used paths (subscript u) and active link capacity constraints (subscript a), then

$$\mathbf{g}_u - \mathbf{B}_u^T\mathbf{z}^* + \mathbf{A}_{ua}^T\mathbf{u}_a^* = \mathbf{0}$$

The active capacity constraints are linearly independent, so they may be determined as follows (see Chapter 3):

$$\mathbf{u}_a^* = (\mathbf{A}_{ua}\mathbf{A}_{ua}^T)^{-1}\mathbf{A}_{ua}(\mathbf{B}_u^T\mathbf{z}^* - \mathbf{g}_u)$$

If $(\mathbf{A}_{ua}\mathbf{A}_{ua}^T)^{-1}$ turns out to be singular, then the constraints are not linearly independent.

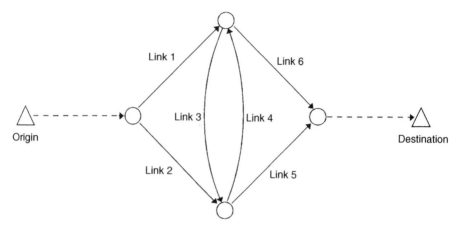

Figure 5.5 *Example network with six links*

As an example, consider the network shown in Fig. 5.5. There is one origin–destination pair, six links, four nodes and four paths. The costs and capacities of the links are shown in Table 5.3.

The total demand from the origin to the destination is 4 units of flow (say, trips per second). The least cost path uses links 1, 3 and 5, but has a capacity of only 1 unit of flow (the capacity of link 1). The next best path uses links 2 and 5, which has a capacity of 1 unit of flow after the best path has been filled. The third path, using links 2, 4 and 6, takes the two remaining units of flow. There is a fourth path using links 1 and 6 which is not used. The link–path incidence matrix, after the deletion of the unused path, is

$$\mathbf{A}_u = \begin{bmatrix} 1 & 0 & 0 \\ 0 & 1 & 1 \\ 1 & 0 & 0 \\ 0 & 0 & 1 \\ 1 & 1 & 0 \\ 0 & 0 & 1 \end{bmatrix}$$

The path costs *exclusive* and *inclusive* of user equilibrium delay are respectively

$$\mathbf{g}_u = \begin{bmatrix} 3 \\ 4 \\ 9 \end{bmatrix} \text{ and } \mathbf{B}_u{}^T \mathbf{z}^* = \begin{bmatrix} 9 \\ 9 \\ 9 \end{bmatrix}$$

Capacity constraints are active on links 1 and 5. When the rows corresponding to all other links are deleted, the following further reduced link–path incidence matrix is

Table 5.3 *Link cost, capacity and flow*

Link	Cost (£/trip)	Capacity (trips/s)	Flow (trips/s)
1	1	1	1
2	3	4	3
3	1	2	1
4	3	4	2
5	1	2	2
6	3	4	2

obtained:

$$\mathbf{A}_{ua} = \begin{bmatrix} 1 & 0 & 0 \\ 1 & 1 & 0 \end{bmatrix}$$

From this, the following matrix product is obtained:

$$\mathbf{A}_{ua}\mathbf{A}_{ua}{}^{T} = \begin{bmatrix} 1 & 1 \\ 1 & 2 \end{bmatrix}$$

This is non-singular, with the following inverse:

$$(\mathbf{A}_{ua}\mathbf{A}_{ua}{}^{T})^{-1} = \begin{bmatrix} 2 & -1 \\ -1 & 1 \end{bmatrix}$$

This yields

$$(\mathbf{A}_{ua}\mathbf{A}_{ua}{}^{T})^{-1}\mathbf{A}_{ua} = \begin{bmatrix} 1 & -1 & 0 \\ 0 & 1 & 0 \end{bmatrix}$$

Finally the dual variables are

$$\mathbf{u}_a{}^* = (\mathbf{A}_{ua}\mathbf{A}_{ua}{}^{T})^{-1}\mathbf{A}_{ua}(\mathbf{B}_u{}^{T}\mathbf{z}^* - \mathbf{g}_u{}^*) = \begin{bmatrix} 1 \\ 5 \end{bmatrix}$$

It may be verified readily that when link 1 has a delay of 1 cost unit and link 5 a delay of 5 cost units, all used paths have equal costs.

5.4 SENSITIVITY

In the preceding section, we hypothesised the existence of a convex function such that the following constrained optimisation problem:

$$\text{Minimum } f(\mathbf{v}) \text{ subject to } \mathbf{v} = \mathbf{Ah}, \mathbf{t} = \mathbf{Bh}, \mathbf{h} \geqslant \mathbf{0}$$

would yield a deterministic user equilibrium assignment. While the existence and uniqueness of a deterministic user equilibrium assignment is not dependent on the existence of such a function (the positive definiteness of the Jacobian of the link costs is sufficient), the finding of the assignment in practice is.

It is not at this stage clear whether the sensitivity expression presented in Chapter 3 is directly applicable, because the constraints are in terms of path flows not link flows. The first constraint can be integrated into the objective function to give

$$\text{Minimum } f(\mathbf{v} = \mathbf{Ah}) \text{ subject to } \mathbf{t} = \mathbf{Bh}, \mathbf{h} \geqslant \mathbf{0}$$

but then the objective function is not strictly convex because there is in general more than one vector $\mathbf{h}$ that yields $\mathbf{v}$. This means that the Hessian of the objective function with respect to $\mathbf{h}$, namely $\nabla^2 f(\mathbf{h})$, is singular and therefore not invertible, as would be required to apply the sensitivity results of Chapter 3 directly.

Assuming for the moment that the non-negativity constraints are inactive at the solution (or, in other words, that all paths are used), then in the absence of link capacity constraints the dual variables are given by

$$\nabla f(\mathbf{h}^*) = \mathbf{B}^{\mathrm{T}}\mathbf{z}^* = \mathbf{g}^*$$

where $\mathbf{z}^*$ is the vector of dual variables relating to the trip table constraint, at the optimum. Note that

$$\nabla f(\mathbf{h}^*) = \mathbf{A}^{\mathrm{T}}\nabla f(\mathbf{v}^*)$$

and that

$$\nabla^2 f(\mathbf{h}^*) = \mathbf{A}^{\mathrm{T}}\nabla^2 f(\mathbf{v}^*)\mathbf{A}$$

Let

$$H^* = \{\mathbf{h}|\mathbf{v}^* = \mathbf{A}\mathbf{h}, \mathbf{t} = \mathbf{B}\mathbf{h}, \mathbf{h} \geqslant \mathbf{0}\}$$

be the set of path flow vectors that are consistent with the uniquely given vector of equilibrium link flows $\mathbf{v}^*$. An *extreme point* of this set is one where a maximum number of path flows are zero. The minimum number of non-zero path flows is given by the rank of $[\mathbf{A}^{\mathrm{T}}, \mathbf{B}^{\mathrm{T}}]^{\mathrm{T}}$. There must be at least one positive path flow for each trip table element that carries a positive demand. Further, the minimum number of non-zero paths cannot exceed the number of linearly independent link flows, which is equal to the total number of links minus the total number of internal nodes (those not connected to a centroid). The Frank–Wolfe algorithm set out earlier usually builds more than the minimum number of paths, so a linear programing problem needs to be solved to find an extreme point.

When an extreme point has been found, the unused paths are deleted from the link–path incidence matrix $\mathbf{A}$ to yield $\mathbf{A}_u$ (subscript u implies 'used') and the matrix $\mathbf{A}_u{}^{\mathrm{T}}\nabla^2 f(\mathbf{v}^*)\mathbf{A}_u$ becomes invertible. Then by making the appropriate substitutions into the sensitivity expression derived in Chapter 3, the following is obtained:

$$d\mathbf{h}_u{}^*/d\mathbf{t} = (\mathbf{A}_u{}^{\mathrm{T}}\nabla^2 f(\mathbf{v}^*)\mathbf{A}_u)^{-1}\mathbf{B}_u{}^{\mathrm{T}}(\mathbf{B}_u(\mathbf{A}_u{}^{\mathrm{T}}\nabla^2 f(\mathbf{v}^*)\mathbf{A}_u)^{-1}\mathbf{B}_u{}^{\mathrm{T}})^{-1}$$

where $\mathbf{B}_u$ is the corresponding reduced origin–destination–path matrix. The invertibility of $\mathbf{A}_u{}^{\mathrm{T}}\nabla^2 f(\mathbf{v}^*)\mathbf{A}_u$ requires that, for a given minimal set of paths, there are *not* two different vectors of paths flows, both compatible with the trip table, that give the *same* (unique) deterministic user equilibrium vector of link flows. If there are two such vectors, then the columns of $\mathbf{A}_u$ are linearly dependent and one path can be eliminated. It follows that if the minimum number of paths have been generated, $\mathbf{A}_u{}^{\mathrm{T}}\nabla^2 f(\mathbf{v}^*)\mathbf{A}_u$ is invertible. Tobin and Friesz (1988) show that it does not matter *which* minimum path set is generated, the sensitivities remain the same.

Note that at the optimum of the equivalent optimisation problem

$$\mathbf{g}_u{}^* = \nabla f(\mathbf{h}^*)_u$$

and

$$\mathbf{g}_u{}^* = \mathbf{B}_u{}^T\mathbf{z}^*$$

The Tobin and Friesz (1988) approach to the analysis of sensitivities extends to the consideration of sensitivities to costs as well. Note that

$$d\mathbf{g} = (\partial\mathbf{g}/\partial\mathbf{h})d\mathbf{h} + (\partial\mathbf{g}/\partial\mathbf{z})d\mathbf{z} = \nabla^2 f(\mathbf{h})d\mathbf{h} + \mathbf{B}^{\mathrm{T}}d\mathbf{z}$$

$$d\mathbf{t} = \mathbf{B}d\mathbf{h}$$

Thus at optimality, the following relationships hold for small changes of the reduced primal and the dual variables, respectively $\Delta\mathbf{h}_u$ and $\Delta\mathbf{z}$:

$$\Delta\mathbf{g}_u^* = \nabla^2 f(\mathbf{h}^*)_u \Delta\mathbf{h}_u + \mathbf{B}_u^T \Delta\mathbf{z}$$

$$\Delta\mathbf{t} = \mathbf{B}_u \Delta\mathbf{h}_u$$

where $\nabla^2 f(\mathbf{h}^*)_u$ is the reduced Hessian. Hence

$$\begin{bmatrix} \Delta\mathbf{g}_u^* \\ \Delta\mathbf{t} \end{bmatrix} = \begin{bmatrix} \nabla^2 f(\mathbf{h}^*)_u & \mathbf{B}_u^T \\ \mathbf{B}_u & \mathbf{0} \end{bmatrix} \begin{bmatrix} \Delta\mathbf{h}_u \\ \Delta\mathbf{z} \end{bmatrix}$$

is the *first order* (or linear) effect of perturbations in the primal and dual variables on the path costs and the origin–destination demands. The matrix between the two perturbation vectors is referred to as the *Jacobian of the system*. The inverse of this Jacobian provides the required sensitivities. Writing the inverse Jacobian in the corresponding block form

$$\begin{bmatrix} \Delta\mathbf{h}_u^* \\ \Delta\mathbf{z}^* \end{bmatrix} = \begin{bmatrix} \mathbf{J}_{11} & \mathbf{J}_{12} \\ \mathbf{J}_{21} & \mathbf{J}_{22} \end{bmatrix} \begin{bmatrix} \Delta\mathbf{g}_u \\ \Delta\mathbf{t} \end{bmatrix}$$

it can be shown that

$$\mathbf{J}_{11} = (\nabla^2 f(\mathbf{h}^*)_u)^{-1}(\mathbf{I} - \mathbf{B}_u^T(\mathbf{B}_u(\nabla^2 f(\mathbf{h}^*)_u)^{-1}\mathbf{B}_u^T)^{-1}\mathbf{B}_u(\nabla^2 f(\mathbf{h}^*)_u)^{-1})$$

$$\mathbf{J}_{12} = (\nabla^2 f(\mathbf{h}^*)_u)^{-1}\mathbf{B}_u^T(\mathbf{B}_u(\nabla^2 f(\mathbf{h}^*)_u)^{-1}\mathbf{B}_u^T)^{-1}$$

$$\mathbf{J}_{21} = (\mathbf{B}_u(\nabla^2 f(\mathbf{h}^*)_u)^{-1}\mathbf{B}_u^T)^{-1}\mathbf{B}_u(\nabla^2 f(\mathbf{h}^*)_u)^{-1}$$

$$\mathbf{J}_{22} = -(\mathbf{B}_u(\nabla^2 f(\mathbf{h}^*)_u)^{-1}\mathbf{B}_u^T)^{-1}$$

When the perturbations in path flows are converted into perturbations in link flows, and when $\nabla^2 f(\mathbf{h}^*)_u$ is replaced by $\mathbf{A}_u^T \nabla^2 f(\mathbf{v}^*)\mathbf{A}_u$, it can be shown that $\mathbf{J}_{12}$ gives the sensitivities obtained earlier. Matrix $\mathbf{J}_{11}$ gives the sensitivity of path flows to perturbations in path costs. If the link cost functions are asymmetric so that $f(\mathbf{v})$ does not exist, the sensitivity expressions may still be used by replacing $\mathbf{A}_u^T \nabla^2 f(\mathbf{v}^*)\mathbf{A}_u$ by $\mathbf{A}_u^T \nabla \mathbf{c}(\mathbf{v}^*)\mathbf{A}_u$.

For some networks, the number of origin–destination pairs exceeds the number of links. For example, 100 zones would lead to 9900 origin–destination pairs (excluding the intrazonal movements). Assuming movement between all origin–destination pairs, this implies *at least* 9900 positive path flows under deterministic user equilibrium assignment. To be able to use the Tobin and Friesz sensitivity expression, there would need to be around 100 links per zone, which is rather more than one might find in practice. If the number of paths included in $\mathbf{A}$ exceeds the number of links, then $\mathbf{A}^T\nabla^2 f(\mathbf{v}^*)\mathbf{A}$ is not invertible because its rank would be insufficient (the rank of the product of matrices is the minimum of the rank of the matrices taken individually).

The sensitivity expression is in any case of limited usefulness because of the dimensions of the matrices to be handled for networks of a practical size. The expression cannot be used with right angle link cost functions, because for such link cost functions, $\nabla^2 f(\mathbf{v}^*) = \mathbf{0}$. The sensitivity expression does, however, offer the interesting insight that for monotonically increasing link cost functions (in other words, for $\nabla \mathbf{c}(\mathbf{v}^*)$ positive definite) link flows change smoothly with changes in demand under deterministic user equilibrium. Furthermore, for networks where the expression is practical, it offers a way of exploring the effect on link flows of small changes in demand. Using the method described in Chapter 3, approximate variances and covariances for link flows may be calculated from variances and covariances for the origin–destination demands. The expression also has a role in bi-level programming, described later in the context of trip table estimation and network design.

Consider the example network shown in Fig. 5.6. This has 1 origin–destination pair, 8 links, 5 internal nodes and 4 paths in total. At most 3 of the paths are linearly independent so at least one of the paths can be omitted to obtain an extreme point. Assume that each link i has the same separable, monotonically increasing cost function of the Bureau of Public Roads type:

$$c_j(v_j) = 6 + v_j{}^4$$

The complete link–path incidence matrix is shown in Table 5.4.

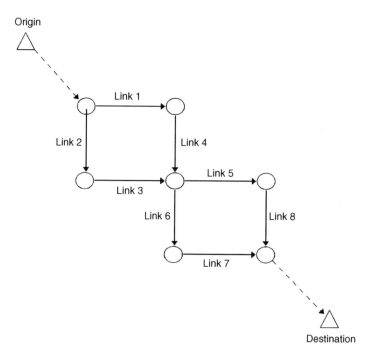

Figure 5.6 *Example network with monotonic cost functions*

Table 5.4 *Link–path incidence matrix pertaining to Fig. 5.6*

Path/Link	1	2	3	4
1	1	0	1	0
2	0	1	0	1
3	0	1	0	1
4	1	0	1	0
5	1	0	0	1
6	0	1	1	0
7	0	1	1	0
8	1	0	0	1

Note that path 1 plus path 2 minus path 3 gives path 4, so one of the columns (say column 4) is linearly dependent. Path 4 is dropped. In fact, path 3 can be dropped as well in this case, because of the symmetry of the costs. The cost on the two remaining used paths must be equal. Suppose that the total demand is 2 units of flow (perhaps vehicles per second). As a result of the symmetry, 1 unit of flow will use path 1 and 1 unit path 2. The cost on paths 1 and 2 will each be 28 units.

For this example, the following are obtained:

$$\mathbf{A_u}^\mathrm{T} \nabla^2 f(\mathbf{v^*}) \mathbf{A_u} = \begin{bmatrix} 16 & 0 \\ 0 & 16 \end{bmatrix}$$

$$(\mathbf{B_u}(\mathbf{A_u}^\mathrm{T} \nabla^2 f(\mathbf{v^*}) \mathbf{A_u})^{-1} \mathbf{B_u}^\mathrm{T})^{-1} = 8$$

This leads to the following sensitivities for path flows to the origin–destination demand

$$\mathbf{J}_{12} = (\mathbf{A_u}^\mathrm{T} \nabla^2 f(\mathbf{v^*}) \mathbf{A_u})^{-1} \mathbf{B_u}^\mathrm{T} (\mathbf{B_u}(\mathbf{A_u}^\mathrm{T} \nabla^2 f(\mathbf{v^*}) \mathbf{A_u})^{-1} \mathbf{B_u}^\mathrm{T})^{-1} = \begin{bmatrix} \frac{1}{2} \frac{1}{2} \end{bmatrix}^\mathrm{T}$$

which suggests that a unit change in demand is equally split between the two paths, as one would expect given the symmetry of the example. A unit increase in the cost of path 1 causes the minimum path cost to increase by half a unit because of the reassignment of traffic to reestablish equilibrium

$$\mathbf{J}_{21} = \begin{bmatrix} \frac{1}{2} \frac{1}{2} \end{bmatrix}$$

Due to linearity (this is only a first order approximation) and the symmetry of the example network, the minimum path cost changes by half the original perturbation in cost, at which point the two path costs are again equal (approximately). The effect of a perturbation of path cost on path flows is given by the following matrix:

$$\mathbf{J}_{11} = \begin{bmatrix} \frac{1}{32} & -\frac{1}{32} \\ -\frac{1}{32} & \frac{1}{32} \end{bmatrix}$$

The interpretation of this is that a unit increase in cost on path 1 would be associated initially with a switch of $\frac{1}{16}$ unit of flow from path 2 to path 1, but to restore equilibrium one half of this, or $\frac{1}{32}$ unit of flow, switches back to path 2. Finally, a perturbation of the origin–destination flow by one unit causes the equilibrium cost to change by 8 units since

$$\mathbf{J}_{22} = 8$$

This may be verified directly by consideration of the link cost functions and by splitting the increase equally between the two paths.

As a second example of sensitivities, consider the network in Fig. 7.1. Each link has the same cost function, namely

$$\mathrm{cost} = 1 + (\mathrm{flow})^2$$

The origin–destination demands from A to C, A to D, B to C and B to D are assumed to be 10 flow units (say vehicles per second) in each case, as before. As a result of the symmetry of the problem, the equilibrium link flows are 20, 20, 5, 5, 5 and 5 units for links 1, 2, 3, 4, 5 and 6 respectively. The path with links 3, 2 and 6, as well as the path with links 4, 1 and

Table 5.5 *Link–origin–destination sensitivities*

Link	Origin–destination pair			
	A to C	A to D	B to C	B to D
1	0.8298	0.5008	0.5008	0.1706
2	0.1702	0.4992	0.4992	0.8294
3	0.0851	0.4991	0.0001	−0.0853
4	−0.0851	−0.0001	0.5009	0.0853
5	−0.0851	0.5009	−0.0001	0.0853
6	0.0851	0.0001	0.4991	−0.0853

5, would not be used, and so are omitted from the link–path and origin–destination–path incidence matrices, namely $\mathbf{A_u}$ and $\mathbf{B_u}$ respectively. The consequence of *not* omitting these paths is that the matrix $\mathbf{A}^T \nabla^2 f(\mathbf{v}^*) \mathbf{A}$ would be singular and therefore not invertible.

Table 5.5 shows the matrix of sensitivities of link flows to elements of the trip table. In relation to links 1 and 2, the symmetry in response is evident. A small increase in the demand for travel from A to C causes reductions in flow on links 4 and 5, because of the increased cost of link 1. Likewise, a small increase in the demand for travel between B and D causes reductions in flow on links 3 and 6, because of the increased cost of link 2.

5.5 MOST LIKELY PATH FLOWS

Except where the paths are limited to a minimal set (referred to above as an *extreme point*), path flows will not be uniquely identified under deterministic user equilibrium assignment. The point has already been made, however, that traffic engineers and transportation planners frequently require estimates of path flows and turning movements. Given a deterministic user equilibrium assignment, there is then a problem of estimating the most likely path flows.

Under deterministic user equilibrium, the cost of all paths used between a particular origin and a particular destination pair will be equal and all paths with costs exceeding this will not be used. Therefore in principle trip-makers will be indifferent as to *which* of the set of paths used they actually choose. For a given origin–destination pair k, the number of permutations giving rise to a given set of path flows is

$$E_k = t_k! / \prod_{j \text{ in } U(k)} h_j!$$

where $U(k)$ is the set of used paths serving origin–destination pair k. Since the trip-makers are indifferent as to which path in $U(k)$ they use, each permutation is equally likely. Across all origin–destination pairs, the number of permutations is given by

$$E(\mathbf{h}) = \prod_k (t_k! / \prod_{j \text{ in } U(k)} h_j!)$$

The number of permutations is frequently referred to as the *entropy* of the system.

Given more than a minimal set of paths, the following entropy maximising problem can be expected to deliver the most likely set of path flows

$$\text{Maximise } E(\mathbf{h}) \text{ subject to } \mathbf{v}^* = \mathbf{Ah}, \, \mathbf{h} \geqslant \mathbf{0}$$

where $\mathbf{v}^*$ is the vector of deterministic user equilibrium link flows and $\mathbf{h}$ is the vector of flows on used paths.

Note that

$$\ln(E(\mathbf{h})) = \sum_k \ln(t_k!) - \sum_k \sum_{j \text{ in } U(k)} \ln(h_j!)$$

Since $E(\mathbf{h})$ is monotonically related to $\ln(E(\mathbf{h}))$, meaning that if $E(\mathbf{h})$ increases then so does $\ln(E(\mathbf{h}))$ and vice versa, and noting that the first term is constant because the trip table is given, the entropy maximising problem may be restated as

$$\text{Minimise } \sum_k \sum_{j \text{ in } U(k)} \ln(h_j!) \text{ subject to } \mathbf{v}^* = \mathbf{Ah}, \mathbf{h} \geqslant \mathbf{0}$$

The use of Sterling's Approximation leads to the following approximately equivalent problem.

$$\text{Minimise } \mathbf{h}^{\mathrm{T}}(\ln(\mathbf{h}) - \mathbf{1}) \text{ subject to } \mathbf{v}^* = \mathbf{Ah}, \mathbf{h} \geqslant \mathbf{0}$$

where $\mathbf{1}$ is a unit vector of appropriate dimension. For practical purposes, the inequality constraints may be neglected because of the form of the objective function, so that at the optimum

$$\ln(\mathbf{h}^*) = \mathbf{A}^{\mathrm{T}}\mathbf{u}$$

where $\mathbf{u}$ is a vector of dual variables (one per link). The solution of this problem by iterative balancing is discussed in Chapter 3.

Consider again the example presented in Section 5.4. The network is shown in Fig. 5.6. Each link has the same, separable cost function, so by symmetry each link carries the same traffic at equilibrium. There are 8 links and 5 internal nodes, so only 3 links are linearly independent. To make u identifiable, only linearly independent constraints are admissible (see Chapter 3). Links 1, 2 and 7 are nominated, but this set of three is not unique. For example, links 3, 4 and 5 would have been another possibility. The dual variables are non-zero only for the nominated three links, so

$$\ln(h_1) = u_1$$
$$\ln(h_2) = u_2 + u_7$$
$$\ln(h_3) = u_1 + u_7$$
$$\ln(h_4) = u_2$$

and for the remaining dual variables

$$u_3 = u_4 = u_5 = u_6 = u_8 = 0$$

As noted, the equilibrium flow on each link is the same, so

$$\exp(u_1) + \exp(u_1 + u_7) = \exp(u_2) + \exp(u_2 + u_7) = \exp(u_1 + u_7) + \exp(u_2 + u_7)$$

This implies that $u_1 = u_2$ and that $u_7 = 0$, which in turn implies that

$$h_1 = h_2 = h_3 = h_4$$

Hence, in the case of this example, the effect of maximising entropy is to *spread* the flow evenly across the four paths. The result is as one would expect, given that each path has two links and all links have the same cost function. Paths are therefore interchangeable.

In practice, the Frank–Wolfe algorithm for finding a deterministic user equilibrium assignment does not deliver a minimum set of paths, so the above-described maximum entropy maximising method is of practical usefulness.

5.6 ELASTIC DEMAND

The deterministic user equilibrium assignment model as presented so far has taken as input a fixed trip table. Concern has been shown recently about the use of fixed trip tables when assessing the traffic consequences of major infrastructure projects, because of the effect such projects can have on the generation and distribution of traffic. The new traffic resulting from such projects is often referred to collectively as *induced traffic*. One way of allowing for this phenomenon in the deterministic user equilibrium assignment model is to let the number of trips between each origin and destination be a function of the cost of travel between the origin and the destination (see Gartner, 1980a,b).

Let the *demand function* be

$$t = t(z)$$

where z is the vector of origin-to-destination *expected costs* for travel. The expected cost is that which the trip-maker would experience *having already made decisions about path choice*. In this context, each trip-maker would seek his best path. One would expect $t(z)$ to be monotonically decreasing, namely

$$(t' - t'')^T(z' - z'') < 0 \qquad \text{for } z' \neq z''$$

When the cost and demand functions are *separable* (namely, when the cost on a link depends only on the flow on that link and the demand between any origin–destination pair depends only on the cost of travel between that origin–destination pair), there is an equivalent optimisation problem that gives rise to a deterministic user equilibrium assignment.

Consider the following objective function

$$f(v, t) = \sum_{i \in I} \int_{x=0}^{x=v_i} c_i(x)\, dx - \sum_{w \in W} \int_{y=0}^{y=t_w} t_w^{-1}(y)\, dy$$

where I is the set of links, W is the set of elements in the trip table and $t_w^{-1}(y)$ is the inverse of the demand function, namely the expected cost of travel between origin–destination pair w. The demand function is invertible because it is monotonically increasing. Consider the following optimisation problem:

Minimise $f(v, t)$ subject to $v = Ah$, $t = Bh$, $h \geqslant 0$, $v \geqslant 0$, $t \geqslant 0$,

where A and B are the link–path and origin–destination–path incidence matrices respectively. It can be shown (see below) that the solution is a deterministic user equilibrium assignment with elastic demand.

When the link cost functions are also monotonically increasing, the objective function has a positive definite Hessian with respect to the primal variables, $\mathbf{v}$ and $\mathbf{t}$, and so is convex. The constraints on the primal variables are also convex. The problem therefore has a unique solution in terms of the primal variables. Note that the optimal primal variables $\mathbf{v}^*$ and $\mathbf{t}^*$ can be uniquely constructed from any set of optimal path flows $\mathbf{h}^*$, even though the optimal path flows themselves will in general not be unique.

The optimality conditions for the path flows $\mathbf{h}^*$ take the form of the following complementary slackness conditions:

$$\mathbf{g}(\mathbf{h}^*) - \mathbf{B}^T\mathbf{z}^* \geqslant \mathbf{0}$$

$$\mathbf{h}^* \geqslant \mathbf{0}$$

$$\mathbf{h}^{*T}(\mathbf{g}(\mathbf{h}^*) - \mathbf{B}^T\mathbf{z}^*) = 0$$

where $\mathbf{g}^*$ is a vector of expected prices with elements given by the inverse of the demand function $g^w_* = t_w^{-1}(t_w)$. The optimality conditions state that if a path is used then its cost is equal to the expected cost of travel from the relevant origin to the relevant destination, and that if the path cost is greater than this the path is not used.

Demand functions that are frequently encountered in the literature have an exponential form

$$t_w = t_0 \exp(-\beta z^w_*)$$

or a linear form

$$t_w = t_0 - \beta z^w_*$$

where t_0 is the maximum demand and β is the sensitivity. The exponential demand function fits conveniently with the logit choice model.

The method of successive averages may be applied as follows:

Method of successive averages algorithm for DUE assignment with elastic demand

Step 1   (initialisation)
$\qquad$ $\mathbf{v} \leftarrow \mathbf{0}$
$\qquad$ $\mathbf{c} \leftarrow \mathbf{c}(\mathbf{v})$
$\qquad$ $n \leftarrow 1$

Step 2   (method of successive averages)
$\qquad$ repeat
$\qquad\qquad$ for link costs $\mathbf{c}$, find least cost paths and their costs $\mathbf{z}$
$\qquad\qquad$ load $\mathbf{t}(\mathbf{z})$ on to least cost paths to get auxiliary link
$\qquad\qquad$ flows $\mathbf{v}^*$
$\qquad\qquad$ $\mathbf{v} \leftarrow (1/n)\mathbf{v}^* + (1 - 1/n)\mathbf{v}$
$\qquad\qquad$ $\mathbf{c} \leftarrow \mathbf{c}(\mathbf{v})$
$\qquad\qquad$ $n \leftarrow n+1$
$\qquad$ until convergence.

As an example, consider the network in Fig. 5.7. The origin-to-destination demand functions are all assumed to have the same form, namely

$$t_{mn} = 10\exp(-0.1z^*_{mn})$$

The maximum demand is 10 units, which decreases as the origin-to-destination cost increases, tending to zero in the limit. Linear cost–flow functions are assumed, with

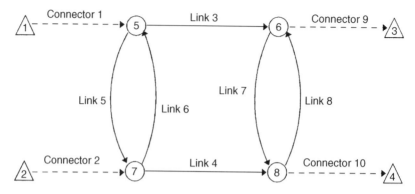

Figure 5.7 *Example network with 8 nodes/centroids*

the intercept and slope terms given in Table 5.6. The fitted link flows and link costs are also given in Table 5.6. The fitted trip table is shown in Table 5.7, together with the respective origin-to-destination costs.

A particular case of elastic demand that has received significant attention in the past is the case of the combined trip distribution and assignment model. The origin and destination totals (the row and column sums of the trip table) are treated as fixed, but the matching of the trip destinations with the trip origins is allowed to be influenced by network conditions. Trip-makers are assumed to choose routes according to the deterministic user equilibrium principle. The model was originally explored by Evans

Table 5.6 *Link cost functions, fitted flows and fitted cost*

Link/ Connector	Origin node	Destination node	Slope	Intercept	Fitted flow	Fitted cost
1	1	5	0.2	2	5.783	3.157
2	2	7	0.2	2	5.786	3.157
3	5	6	0.5	2	5.787	4.894
4	7	8	0.5	2	5.781	4.891
5	5	7	1	2	0.594	2.600
6	7	5	1	2	0.599	2.605
7	6	8	0.3	2	1.949	2.583
8	8	6	0.3	2	1.944	2.582
9	6	3	0.2	2	5.783	3.157
10	8	4	0.2	2	5.786	3.157

Table 5.7 *Origin-to-destination fitted flows and expected costs*

Origin	Destination	Fitted flow	Expected cost
1	3	3.261	11.207
1	4	2.518	13.790
2	3	2.519	13.786
2	4	3.261	11.205

(1976), who proposed an equivalent optimisation problem based on the entropy derivation of the gravity model.

The number of permutations of trips that give rise to a given trip table $\mathbf{t}$ is

$$E(\mathbf{t}) = \left(\sum_k t_k\right)! \Big/ \prod_k t_k!$$

This is frequently referred to as the *entropy* of the trip table. It has been argued (see Wilson, 1967) that the trip table with maximum entropy is the trip table most likely to arise in reality. Note that

$$\ln(E(\mathbf{t})) = \ln\left(\left(\sum_k t_k\right)!\right) - \sum_k \ln(t_k!)$$

Since $E(\mathbf{t})$ is monotonically related to $\ln(E(\mathbf{t}))$, meaning that if $E(\mathbf{t})$ increases then so does $\ln(E(\mathbf{t}))$ and vice versa, and noting that the first term is constant because the total number of trips in the trip table is specified by the row and column totals, the maximisation of $E(\mathbf{t})$ is equivalent to the minimisation of $\sum_k \ln(t_k!)$. Sterling's approximation suggests that

$$\sum_k \ln(t_k!) \approx \mathbf{t}^{\mathrm{T}}(\ln(\mathbf{t}) - \mathbf{1})$$

Combining the above entropy-related function with the sum of the integrals of the link cost functions yields the objective function considered by Evans (1976), namely

$$f(\mathbf{v}, \mathbf{t}) = \alpha \sum_{i=1 \text{ to } N} \int_{x=0}^{x=v_i} c_i(x)\,\mathrm{d}x + \mathbf{t}^{\mathrm{T}}(\ln(\mathbf{t}) - \mathbf{1})$$

where α is the parameter of the gravity model deterrence function. Note that

$$\mathbf{v} = \mathbf{Ah}$$

and

$$\mathbf{t} = \mathbf{Bh}$$

so the objective function may be expressed solely in terms of path flows. Let $O(i)$ be the set of paths with origin i and $D(j)$ the set of paths with destination j. Then the equivalent optimisation problem may be written as

$$\text{Minimise } f(\mathbf{h}) \text{ subject to } o_m = \sum_{j \text{ in } o(m)} h_j, \; d_n = \sum_{j \text{ in } D(n)} h_j, \; \mathbf{h} \geqslant \mathbf{0}$$

where o_m is the total number of trips generated by origin m and d_n is the total number of trips attracted by destination n. The Lagrangian equation has the form

$$L(\mathbf{h}, \gamma, \eta) = f(\mathbf{h}) - \sum_m \gamma_m \left(o_m - \sum_{j \text{ in } O(m)} h_j\right) - \sum_n \phi_n \left(d_n - \sum_{j \text{ in } D(n)} h_j\right)$$

where γ_m is a dual variable for origin m and ϕ_n is a dual variable for destination n. The Kuhn–Tucker conditions are

$$\alpha g_j + \ln(t_k) + \gamma_m + \phi_n \geqslant 0$$
$$h_j \geqslant 0$$
$$(\alpha g_j + \ln(t_k) + \gamma_m + \phi_n)h_j = 0$$

where path j connects origin m to destination n, which in turn constitutes trip table element k, and where g_j is the cost of path j. Let z_k be the expected cost of travel from origin m to destination n. Then the resulting gravity model has the form

$$t_k{}^* = \exp(-\alpha z_k - \gamma_m - \phi_n) = R_m S_n \exp(-\alpha z_k)$$

where

$$R_m = \exp(-\gamma_m)$$

and

$$S_n = \exp(-\phi_n)$$

are the row and column balancing factors respectively. They can be calculated iteratively using the method described in Chapter 3 so that the row and column sums of the fitted trip table agree with the pre-specified totals.

Substituting the gravity model into the complementary slackness conditions yields

$$\alpha(g_j - z_k) \geqslant 0$$
$$h_j \geqslant 0$$
$$\alpha(g_j - z_k)h_j = 0$$

implying that path j will not be used if the path cost is greater than the expected cost of travel for the corresponding trip table element k.

The solution procedure proposed by Evans utilises the fact that the gravity model, combined with an all-or-nothing assignment, provides a descent direction. To demonstrate this, let $\mathbf{h}$ be the vector of current path flows. This vector is assumed to be feasible, in the sense that the path flows agree with pre-specified origin and destination totals and are non-negative. At this point, the gradient of the objective function is

$$\nabla f(\mathbf{h}) = \alpha\mathbf{g} + \mathbf{B}^{\mathrm{T}}\ln(\mathbf{t})$$

where

$$\mathbf{t} = \mathbf{Bh}$$

The gravity model yields

$$\ln(t_k{}^*) = -\alpha z_k - \gamma_m - \phi_n$$

where the dual variables γ_m and ϕ_n are calculated so that the pre-specified row and column totals are conserved. By noting that

$$\ln(\mathbf{t}^*) + \alpha\mathbf{z} + \mathbf{\Gamma}^{\mathrm{T}}\boldsymbol{\gamma} + \mathbf{\Phi}^{\mathrm{T}}\boldsymbol{\phi} = \mathbf{0}$$

where Γ and Φ are origin–path and destination–path incidence matrices respectively, the gradient vector may be re-expressed as

$$\nabla f(\mathbf{h}) = \mathbf{B}^T(\ln(\mathbf{t})) - \ln(\mathbf{t}^*) + (\mathbf{g} - \mathbf{B}^T\mathbf{z})\alpha - \Gamma^T\gamma - \Phi^T\phi$$

The gravity model estimate of the trip table is assigned to the current set of least cost paths (an all-or-nothing assignment) to yield auxiliary path flows $\mathbf{h}^*$. Pre-multiplication of the gradient with the direction provided by the auxiliary path flows yields

$$(\mathbf{h}^* - \mathbf{h})^T\nabla f(\mathbf{h}) = (\mathbf{h}^* - \mathbf{h})^T\mathbf{B}^T(\ln(\mathbf{t}) - \ln(\mathbf{t}^*)) + (\mathbf{h}^* - \mathbf{h})^T(\mathbf{g} - \mathbf{B}^T\mathbf{z})\alpha$$
$$= (\mathbf{t}^* - \mathbf{t})^T(\ln(\mathbf{t}) - \ln(\mathbf{t}^*)) + (\mathbf{h}^* - \mathbf{h})^T(\mathbf{g} - \mathbf{B}^T\mathbf{z})\alpha < 0$$

since the feasibility of $\mathbf{h}$ and $\mathbf{h}^*$ requires that

$$(\mathbf{h}^* - \mathbf{h})^T(\Gamma^T\gamma + \Phi^T\phi) = 0$$

The first term of the inequality is negative because of the concavity of the natural log transformation. The second term is non-positive because $\mathbf{z}$ is the vector of least origin-to-destination path costs and

$$\mathbf{h}^{*T}(\mathbf{g} - \mathbf{B}^T\mathbf{z}) = 0$$

while

$$\mathbf{h}^T(\mathbf{g} - \mathbf{B}^T\mathbf{z}) \geqslant 0$$

Hence $(\mathbf{h}^* - \mathbf{h})$ is a descent direction.

In practice, it is not necessary to specify the paths. Starting with an initial set of empty network minimum origin-to-destination travel costs $\mathbf{z}$, the gravity model is fitted to yield the auxiliary trip table $\mathbf{t}^*$. An all-or-nothing assignment is performed to obtain auxiliary link flows $\mathbf{v}^*$ (the path flows themselves are not actually required), which are then averaged with the current set of link flows. The link costs are recalculated and the minimum origin-to-destination travel costs $\mathbf{z}$ are sought again, etc. This leads to the following algorithm.

Evans algorithm for the combined distribution and assignment problem

Step 1 (initialisation)
   $\mathbf{v} \leftarrow \mathbf{0}$
   $n \leftarrow 0$

Step 2 (search direction)
   $n \leftarrow n + 1$
   $\mathbf{z} \leftarrow$ Find least cost paths based on $\mathbf{c}(\mathbf{v})$
   $\mathbf{t}^* \leftarrow$ Fit gravity model $t_k^* = R_m S_n \exp(-\alpha z_k)$ based on o_m and d_n
   $\mathbf{v}^* \leftarrow$ Assign $\mathbf{t}^*$ to least cost paths

Step 3 (method of successive averages)
   $\mathbf{v} \leftarrow \mathbf{v}^*(1/n) + \mathbf{v}(1 - 1/n)$
   if convergence insufficient then return to Step 2.

As an example, consider the network in Fig. 7.1. Each link has the same cost function, namely

$$\text{cost} = 1 + (\text{flow})^2$$

Table 5.8 *Combined distribution and assignment link flows*

	Links					
Iterations	1	2	3	4	5	6
10	7.6662	7.3338	3.7820	1.4482	3.7962	1.1299
20	7.5930	7.4070	3.3633	0.9564	3.3803	0.7873
50	7.5382	7.4618	3.1242	0.6626	3.1320	0.5938
100	7.5192	7.4808	3.0450	0.5642	3.0490	0.5298
200	7.5096	7.4904	3.0052	0.5148	3.0072	0.4976
400	7.5048	7.4952	2.9926	0.4888	2.9788	0.4822
Cost	57.3221	57.1780	9.9110	1.2400	9.9170	1.2317

Table 5.9 *Combined distribution and assignment trip tables*

	Origin-destination pair			
Link	A to C	A to D	B to C	B to D
10	4.0733	5.9267	0.9267	4.0733
20	4.0686	5.9314	0.9314	4.0686
50	4.0674	5.9326	0.9326	4.0674
100	4.0686	5.9314	0.9314	4.0686
200	4.0695	5.9305	0.9305	4.0695
400	4.0700	5.9300	0.9300	4.0700

The origins A and B generate 10 and 5 units of flow respectively while destinations C and D attract 5 and 10 units of flow respectively. The path with links 3, 2 and 6 as well as the path with links 4, 1 and 5 would not be used (paths with loops are of course also excluded). Table 5.8 gives the link flows obtained from the above Frank–Wolfe algorithm for varying numbers of iterations and the link costs for the 400th iteration. Table 5.9 gives the corresponding fitted trip tables. Convergence is evidently fairly slow.

5.7 TIME-DEPENDENT NETWORKS

The space–time network allows consideration of *time-varying* conditions (see Chapter 3). Nodes represent locations in space and time. Links have constant costs and fixed capacities, as in the previous section, but delay is represented explicitly by links which connect one node in one time interval to itself in a later time interval (see Fig. 2.8 for an example of a space–time network). It is tempting to formulate a linear programming problem with total system cost as the objective, as was done for deterministic user equilibrium assignment with right angle cost functions. If this is done, it does indeed produce an assignment whereby the least cost paths are filled first, followed by the second best paths, etc., as previously. The assignment is therefore a deterministic user equilibrium in the sense that no trip-maker can reduce his travel cost by changing path.

The difficulty is that there is nothing in the problem as just described to ensure that a first-in, first-out (FIFO) discipline is observed. One unit of flow may arrive at a link *before* but leave *after* another unit of flow (as illustrated in Fig. 2.8). In a single commodity context, units of flow are interchangeable and so FIFO has no meaning. However, in a multiple commodity context, for example where there is more than one origin–destination pair, the violation of FIFO poses difficulties. Carey (1992) has shown that FIFO results in inherently non-convex constraints. While $\mathbf{v}_1$ and $\mathbf{v}_2$ may be two FIFO-feasible vectors of link flows *disaggregated by commodity* in a space–time network, the vector $\mathbf{v}_1\lambda + \mathbf{v}_2(1-\lambda)$ is quite likely to be FIFO-infeasible for $1 \geqslant \lambda \geqslant 0$. Various attempts have been made to impose some approximate form of FIFO discipline on space–time networks (see Drissi-Kaitouni and Hameda-Benchekroun, 1992), but none have been very convincing.

There are two pertinent questions. The first is whether FIFO applies in reality. Although queuing processes do impose a FIFO discipline, overtaking can and does occur on many kinds of road. Indeed, if overtaking is motivated by relative cost savings (for example, a competitive regime could be envisaged where A overtakes B if A's cost saving by so doing is greater than B's cost saving if B overtakes A), then the minimisation of total system cost in space–time networks may produce a realistic traffic assignment for road networks.

A second question is what effect random flow variation has. Since the convex combination of two FIFO-feasible link flow vectors may be FIFO-infeasible, the average of a series of FIFO-feasible link flow vectors may also be FIFO-infeasible. While average link flows disaggregated by commodity are likely to be FIFO-infeasible, the pattern of average flows will still be *influenced* by FIFO (if it is a feature of reality), because the averages would be based only on FIFO-feasible patterns. Hence FIFO cannot be ignored entirely.

Noting that queues in one period must be processed in subsequent periods, a potentially useful compromise is the approximation of a dynamic assignment by a sequence of steady state equilibria linked across time by the queues. Equilibrium delay is a cost penalty required to bring demand into line with a fixed supply of infrastructure. However, underlying the notion of delay is some form of queuing process, so a natural extension would be to translate equilibrium delays into equilibrium queues and carry the queues across periods. By explicitly introducing queues, the temporary overloading of links can be modelled.

The equilibrium queue is equal to the equilibrium delay times the service rate. For example, if the equilibrium delay is 10 seconds and the service rate is 0.5 vehicles per second, the equilibrium queue will be 5 vehicles. Let $\mathbf{S}$ be a diagonal matrix of *link service rates*:

$$\mathbf{S} = \begin{bmatrix} s_1 & & \\ & s_2 & \\ & & \ddots \end{bmatrix}$$

then

$$\mathbf{d_t} = \mathbf{S}^{-1}\mathbf{q_t}$$

where $\mathbf{d_t}$ is a vector of equilibrium link delays and $\mathbf{q_t}$ is a vector of equilibrium link queues. Note that the link service rate has been referred to earlier as the link capacity.

The link capacity constraints can be reformulated as the following complementary slackness conditions:

$$\mathbf{s} - \mathbf{Ah}_t - \mathbf{q}_{t-1} + \mathbf{q}_t \geqslant \mathbf{0}$$

$$\mathbf{q}_t \geqslant \mathbf{0}$$

$$(\mathbf{s} - \mathbf{Ah}_t - \mathbf{q}_{t-1} + \mathbf{q}_t)^\mathrm{T}\mathbf{q}_t = 0$$

where subscript t denotes a period, $\mathbf{q}_{t-1}$ is the vector of queues at the end of period $t-1$ and $\mathbf{q}_t$ is the vector of queues at the end of period t. The demand for the links $\mathbf{Ah}_t$ can therefore exceed their steady state capacities $\mathbf{s}$, provided the equilibrium queues grow sufficiently between periods to absorb the excess.

The above complementary slackness conditions capture the macroscopic consequences of FIFO. Queues left over from a preceding period are processed *first*, effectively reducing the link capacities. New queues are allowed to form *only* when the corresponding capacities are exceeded.

The following is a simplified version of the model proposed by Lam *et al.* (1995) for stochastic user equilibrium assignment. Consider the following quadratic objective function:

$$f(\mathbf{h}_t, \mathbf{q}_t) = \mathbf{h}_t{}^{\mathrm{T}}\mathbf{A}^{\mathrm{T}}\mathbf{c} + 0.5\mathbf{q}_t{}^{\mathrm{T}}\mathbf{S}^{-1}\mathbf{q}_t = \mathbf{h}_t{}^{\mathrm{T}}\mathbf{g} + 0.5\mathbf{q}_t{}^{\mathrm{T}}\mathbf{S}^{-1}\mathbf{q}_t$$

which is minimised subject to trip table constraints

$$\mathbf{t}_t = \mathbf{B}\mathbf{h}_t$$

and link capacity constraints

$$\mathbf{s} \geqslant \mathbf{A}\mathbf{h}_t + \mathbf{q}_{t-1} - \mathbf{q}_t$$

The Hessian of $f(\mathbf{h}_t, \mathbf{q}_t)$ is positive semi-definite so the objective function is convex. The constraints are also convex, so there is a (possibly not unique) solution to the minimisation problem. The Lagrangian equation has the form

$$L(\mathbf{h}_t, \mathbf{q}_t, \mathbf{u}_t, \mathbf{z}_t) = f(\mathbf{h}_t, \mathbf{q}_t) + \mathbf{z}_t{}^{\mathrm{T}}(\mathbf{t}_t - \mathbf{B}\mathbf{h}_t) - \mathbf{u}_t{}^{\mathrm{T}}(\mathbf{s} - \mathbf{A}\mathbf{h}_t - \mathbf{q}_{t-1} + \mathbf{q}_t)$$

The optimality conditions for $\mathbf{q}_t$ identify the dual variables $\mathbf{u}_t$ as being equal to the equilibrium delays, since

$$\mathbf{S}^{-1}\mathbf{q}_t{}^* = \mathbf{u}_t = \mathbf{d}_t{}^*$$

The optimality conditions for $\mathbf{h}_t$ are

$$\mathbf{g}_t + \mathbf{A}^{\mathrm{T}}\mathbf{u}_t = \mathbf{B}^{\mathrm{T}}\mathbf{z}_t$$

which, together with the interpretation of $\mathbf{u}_t$ as the equilibrium delays and $\mathbf{z}_t$ as the minimum origin-to-destination travel cost inclusive of delay, yield a deterministic user equilibrium assignment. The optimality conditions for $\mathbf{u}_t$ are

$$\mathbf{s} - \mathbf{A}\mathbf{h}_t - \mathbf{q}_{t-1} + \mathbf{q}_t \geqslant \mathbf{0}$$

$$\mathbf{u}_t{}^* \geqslant \mathbf{0}$$

$$\mathbf{u}_t{}^{*\mathrm{T}}(\mathbf{s} - \mathbf{A}\mathbf{h}_t - \mathbf{q}_{t-1} + \mathbf{q}_t) = 0$$

Noting that a positive element of $\mathbf{u}_t{}^*$ corresponds to a positive element of $\mathbf{q}_t{}^*$ and a zero element of $\mathbf{u}_t{}^*$ corresponds to a zero element of $\mathbf{q}_t{}^*$, these optimality conditions give the required capacity-queue complementary slackness conditions presented earlier.

Although temporary overloading of the network can be accommodated through queue formation, the macroscopic effects of FIFO are reproduced. Since the role of the queue in one period is to reduce the capacity of the link in the subsequent period, the composition of the queue in terms of user class, origin or destination is not relevant. Despite the explicit treatment of queues, the model is essentially steady state, since all the trips in the trip table must be completed within the period. The queues are demand-determined artefacts to bring demand in line with the infrastructure supply.

The model may be implemented by quadratic programming in a recursive way. Starting with q_0, q_1 is estimated, and from this, q_2 may be estimated, etc.

5.8 DISCUSSION

This chapter has concentrated on deterministic user equilibrium assignment, which is perhaps becoming one of the most widely applied assignment principles. Numerous software packages are now available for finding such an assignment. While it is based on convincing assumptions about trip-making behaviour, it also assumes that trip-makers have identical perceptions about the costs involved. This has frequently been interpreted as perfect information or perfect foresight. Whatever the interpretation, the assumption is clearly unrealistic in practice. Furthermore, in the context of traveller information systems, it is unhelpful because one of the central aims of such systems is to reduce the amount of uncertainty regarding the costs of travel.

The deterministic user equilibrium principle extends quite readily to multi-modal networks, provided of course the network has been specified to represent the penalties of modal interchange correctly. In this way, *mode choice* may be included within the framework presented here. Various elastic demand formulations are also possible (two are presented here).

This chapter has concentrated almost entirely on the steady state problem. The extension of constant cost, fixed capacity deterministic user equilibrium to space–time networks has been considered. However, this has been sufficient to illustrate the difficulties, not only in relation to FIFO but also in relation to assumed travel behaviour (in particular, knowledge acquisition). An *ex post* deterministic user equilibrium, such as investigated by Addison and Heydecker (1993), would require trip-makers to have perfect foresight, which is even less realistic than an assumption of the identical perception of costs. An alternative approach, followed by Wie *et al.* (1990) and others, is to assume that path choice decisions are based on the instantaneous costs. This allows decisions to be revised (but not retrospectively) as the trips progress. While removing any requirement for perfect foresight, it also eliminates any allowance for expectations of future costs.

Because of the complexities involved in finding a time-dependent deterministic user equilibrium assignment, many researchers and practitioners have resorted to simulation. An early, but still current, simulation tool of this kind is CONTRAM (Leonard *et al.*, 1978). This may be characterised as a *packet simulator*. Packets of traffic travel down links and are processed at nodes. When the arrivals exceed the departures, packets are queued at the nodes. Each packet is assigned to a minimum cost path. The demand is specified in the form of a series of trip tables, one for each time slice. Link cost is flow-dependent. Later versions of the model include more than one class of traffic, with class-specific rules and link cost functions. The required assignment of packets is sought by an iterative heuristic procedure that has stood the test of time.

A useful compromise approach to dynamic assignment, which makes use of network flow programming methods while taking queues and their effect on link capacity explicitly into account, has been presented. The queues are allowed to accumulate and decline over a number of periods. However, all the trips in the trip table are also assumed to be completed within the period; a contradiction, which is believed not to be serious in practice. The macroscopic effects of FIFO, in particular the precedence of existing queues over new arrivals, are respected. There is as yet no practical experience with the approach.

6

Stochastic User Equilibrium Assignment

6.1 INTRODUCTION

In this chapter, the assumption made in deterministic user equilibrium assignment that trip-makers have perfect information is relaxed. Consider the two link network in Fig. 5.1. The principle of a stochastic user equilibrium assignment is illustrated for monotonically increasing cost functions in Fig. 6.1 and for right angle cost functions in Fig. 6.2. The *demand curve* for the two paths, in relation to the difference in cost between the two paths, is now logistic in shape. Rather than flip-flop path choice behaviour about the point where path costs are equal, there will be some trip-makers who are apparently happy to choose a higher cost path. The *supply curve* gives the relationship between the division of traffic between the two paths and their relative costs. In the case of monotonically increasing link costs, the supply curve is smooth, logistic in shape, but opposite in direction to the demand curve. In the case of right angle cost functions, the supply curve is a step function. Within the feasible range of path shares, the cost difference is constant and the supply curve horizontal. At either extreme the cost difference is undefined and the supply curve vertical. For both types of cost function, the equilibrium is given by the intersection of the demand and supply curves.

If the choice probability for path 1 is considered in relation to the difference in cost between the two paths, there is a smooth *response surface* (see Fig. 6.3); note that the response surface is the demand curve, rotated by 90°. This is not only a more realistic behavioural assumption but also brings practical advantages. One significant practical advantage is that, under a stochastic user equilibrium, path flows are uniquely determined, in contrast to under a deterministic user equilibrium, where they are in general not. When path costs are equal, the step response surface of deterministic user equilibrium assignment does not specify the path share, whereas the smooth response surface of stochastic user equilibrium assignment does.

There are two commonly adopted approaches to stochastic user equilibrium assignment; the *logit* and *probit* path choice models. Both give rise to smooth response surfaces, like that portrayed in Fig. 6.3, but differ in terms of the functional form used to fit the surface. The logit model is more tractable as it offers an equivalent optimisation problem (at least when the Jacobian of the link cost functions is positive definite), but it treats all choice alternatives as if they were statistically independent. The probit model can allow for correlations between alternative paths, but has no known equivalent optimisation problem.

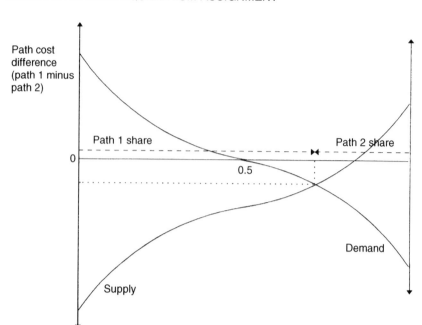

Figure 6.1 *Stochastic user equilibrium assignment for increasing cost functions*

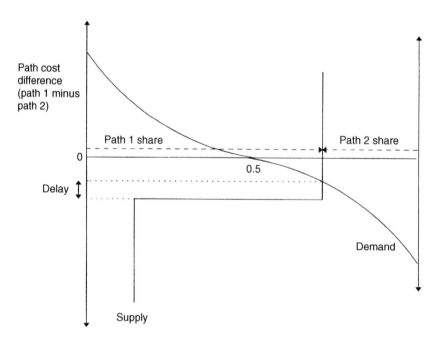

Figure 6.2 *Stochastic user equilibrium assignment for right angle cost functions*

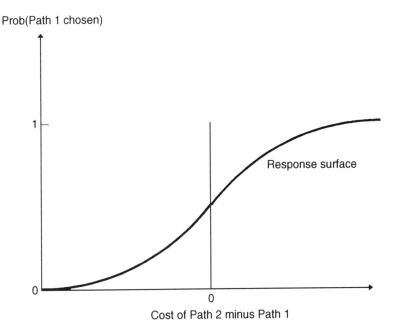

Figure 6.3 *Response surface for stochastic user equilibrium assignment*

6.2 EXISTENCE AND UNIQUENESS

Let **p** be the vector of path choice probabilities with elements p_k giving the probability of selecting path k. Since the probabilities must sum to one across the paths that connect each origin–destination pair

$$1 = \mathbf{Bp}$$

Estimated path flows are arrived at by multiplying the probabilities by the origin–destination demand as follows

$$h_j = p_j t_k$$

where path j connects origin–destination pair k and t_k is the corresponding origin–destination flow (the kth element of the trip table).

Under stochastic user equilibrium assignment, the demand for paths is related to their costs, so

$$\mathbf{h} = \mathbf{h}^D(\mathbf{g})$$

where $\mathbf{h}^D(\mathbf{g})$ is a demand function. In general, it is expected that $dh_j/dg_j < 0$, or in other words, that the demand for path j falls with increasing cost for path j. Of course, the cost for path j will also affect the demand for any other path j', where $j \neq j'$, connecting the same origin to the same destination, but overall it is expected that

$$(\mathbf{g} - \mathbf{g}^*)^{\mathrm{T}}(\mathbf{h}^D(\mathbf{g}) - \mathbf{h}^D(\mathbf{g}^*)) < 0$$

where $\mathbf{g}$ and $\mathbf{g}^*$ are two sets of path costs. This implies that the Jacobian of the demand function $\mathbf{h}^D(\mathbf{g})$ is negative definite.

If the Jacobian of $\mathbf{h}^D(\mathbf{g})$ is negative definite, the demand function may be inverted to yield

$$(\mathbf{h} - \mathbf{h}^*)^{\mathrm{T}}(\mathbf{g}^D(\mathbf{h}) - \mathbf{g}^D(\mathbf{h}^*)) < 0$$

where $\mathbf{g}^D(\mathbf{h})$ represents the **path prices** trip-makers are willing to pay for path flows $\mathbf{h}$. If $\mathbf{h}^D(\mathbf{g})$ is negative definite then so is $\mathbf{g}^D(\mathbf{h})$.

On the supply side, **path costs** are related to path flows. In general, path costs are expected to increase (or at least not decrease) with increasing path flows. The relationship between path costs and path flows is discussed in Chapter 4. In Chapter 5, a distinction was made between the case of monotonically increasing link cost functions and right angle link cost functions; the same distinction will be made here.

Consider first the case of monotonically increasing link cost functions

$$(\mathbf{v} - \mathbf{v}^*)^{\mathrm{T}}(\mathbf{c}(\mathbf{v}) - (\mathbf{c}(\mathbf{v}^*)) > 0$$

This may be converted into path costs and flows as follows:

$$(\mathbf{v} - \mathbf{v}^*)^{\mathrm{T}}(\mathbf{c}(\mathbf{v}) - \mathbf{c}(\mathbf{v}^*)) = (\mathbf{h} - \mathbf{h}^*)^{\mathrm{T}}\mathbf{A}^{\mathrm{T}}(\mathbf{c}(\mathbf{v}) - \mathbf{c}(\mathbf{v}^*))$$

$$= (\mathbf{h} - \mathbf{h}^*)^{\mathrm{T}}(\mathbf{g}^S(\mathbf{h}) - \mathbf{g}^S(\mathbf{h}^*)) \geqslant 0$$

where $\mathbf{g}^S(\mathbf{h})$ corresponds to the path costs at path flow levels $\mathbf{h}$. The inequality loses its strictness, because there are potentially many path flow vectors which yield a given link flow vector and therefore a given link cost vector and a given path cost vector. In other words, it is possible that

$$\mathbf{g}^S(\mathbf{h}) = \mathbf{g}^S(\mathbf{h}^*) \qquad \text{for } \mathbf{h} \neq \mathbf{h}^*$$

Note that the supply function has a positive semi-definite Jacobian, indicating that any increase in path flow tends to be associated with non-decreases in path costs.

Taken together,

$$(\mathbf{h} - \mathbf{h}^*)^{\mathrm{T}}(\mathbf{g}^D(\mathbf{h}) - \mathbf{g}^D(\mathbf{h}^*)) < 0$$

and

$$(\mathbf{h} - \mathbf{h}^*)^{\mathrm{T}}(\mathbf{g}^S(\mathbf{h}) - \mathbf{g}^S(\mathbf{h}^*)) \geqslant 0$$

imply that

$$(\mathbf{h} - \mathbf{h}^*)^{\mathrm{T}}((\mathbf{g}^D(\mathbf{h}) - \mathbf{g}^S(\mathbf{h})) - (\mathbf{g}^D(\mathbf{h}^*) - \mathbf{g}^S(\mathbf{h}^*))) < 0$$

Let

$$\mathbf{d}(\mathbf{h}) = \mathbf{g}^D(\mathbf{h}) - \mathbf{g}^S(\mathbf{h})$$

be the vector of differences between the path prices trip-makers would be willing to pay and the path costs. Then

$$(\mathbf{h} - \mathbf{h}^*)^{\mathrm{T}}(\mathbf{d}(\mathbf{h}) - \mathbf{d}(\mathbf{h}^*)) < 0$$

implying that the Jacobian of $\mathbf{d}(\mathbf{h})$ is negative definite.

Path flows are subject to convex feasibility constraints of the following form:

$$\mathbf{t} = \mathbf{Bh}, \qquad \mathbf{h} \geqslant \mathbf{0}$$

At equilibrium, one of the following relationships is expected between the equilibrium flow on path j (h_j^*), any other feasible flow on path j (h_j), and the difference between the price for and the cost of path j (d_j^*):

$$h_j \geqslant h_j^* = 0 \text{ if } d_j^* \leqslant 0$$

(no flow at equilibrium when the price offered is less than the cost),
or

$$h_j \geqslant 0 \text{ and } h_j^* \geqslant 0 \text{ if } d_j^* = 0$$

(flow at equilibrium greater than or equal to zero when the price offered matches the cost),
or

$$h_j^* \geqslant h_j \geqslant 0 \text{ if } d_j^* \geqslant 0$$

(flow at equilibrium is constrained to be less than a certain level when the price offered exceeds the cost).

Taken together, these conditions may be written as

$$(\mathbf{h} - \mathbf{h}^*)^{\mathrm{T}} \mathbf{d}(\mathbf{h}^*) \leqslant \mathbf{0}$$

As in the case of a deterministic user equilibrium, the *existence* of a stochastic user equilibrium is given under very general conditions relating to the convexity of the constraints.

Taking any vector of price–cost differences $\mathbf{d}$, consider the following linear programming problem

$$\text{Minimise } -\mathbf{h}^{\mathrm{T}} \mathbf{d} \text{ subject to } \mathbf{t} = \mathbf{Bh}, \ \mathbf{h} \geqslant \mathbf{0}$$

This has a solution, provided the set of feasible path flows is non-empty. As both the objective function and the constraints are convex, the optimum path flows $\mathbf{h}^*$ are fully (but not necessarily uniquely) identified by the following inequality

$$(\mathbf{h} - \mathbf{h}^*)^{\mathrm{T}} \mathbf{d} \leqslant 0 \qquad \text{for all feasible } \mathbf{h}$$

(see Chapter 3 on the optimality conditions for convex programming problems). Since we can substitute $\mathbf{d}$ by $\mathbf{d}(\mathbf{h}^*)$, the *existence* of a stochastic user equilibrium assignment is assured, provided of course that the set of feasible path flows is non-empty.

Given the existence of a stochastic user equilibrium assignment, the negative definiteness of $\nabla \mathbf{d}(\mathbf{h})$ is *necessary* and *sufficient* for its uniqueness. Suppose there are two distinct equilibria, $\mathbf{h}^*$ and $\mathbf{h}^{**}$, with $\mathbf{h}^* \neq \mathbf{h}^{**}$. The negative definiteness of the Jacobian of $\mathbf{d}(\mathbf{h})$ ensures that

$$(\mathbf{h}^* - \mathbf{h}^{**})^{\mathrm{T}} (\mathbf{d}(\mathbf{h}^*) - \mathbf{d}(\mathbf{h}^{**})) < 0$$

As both $\mathbf{h}^*$ and $\mathbf{h}^{**}$ are equilibria

$$(\mathbf{h}^{**} - \mathbf{h}^*)^{\mathrm{T}} \mathbf{d}(\mathbf{h}^*) \leqslant 0$$

and

$$(\mathbf{h}^* - \mathbf{h}^{**})^{\mathrm{T}} \mathbf{d}(\mathbf{h}^{**}) \leqslant 0$$

Together, these suggest that

$$(\mathbf{h}^* - \mathbf{h}^{**})^{\mathrm{T}}(\mathbf{d}(\mathbf{h}^*) - \mathbf{d}(\mathbf{h}^{**})) \geqslant 0$$

which contradicts the negative definiteness of the Jacobian of $\mathbf{d}(\mathbf{h})$. *Hence under stochastic user equilibrium assignment the path flows are unique.*

An *equivalent optimisation problem* is of particular interest from the point of view of finding a stochastic user equilibrium assignment in transportation networks. If there is an optimisation problem of the form

$$\text{Minimise } f(\mathbf{h}) \text{ subject to } \mathbf{t} = \mathbf{Bh} \text{ and } \mathbf{h} \geqslant \mathbf{0}$$

then its solution would be the required assignment, provided $f(\mathbf{h})$ were convex with

$$\nabla f(\mathbf{h}) = -\mathbf{d}(\mathbf{h})$$

This is because, at the optimum (see Chapter 3 for the optimality conditions of convex programing problems),

$$(\mathbf{h} - \mathbf{h}^*)^{\mathrm{T}}\mathbf{d}(\mathbf{h}^*) \leqslant 0 \qquad \text{for all feasible } \mathbf{h}$$

and the constraints on $\mathbf{h}$ are, as already noted, convex. If the Jacobian of $\mathbf{d}(\mathbf{h})$ is positive definite, the solution is also unique, as shown above (see also Chapter 3).

Note that $\nabla^2 f(\mathbf{h})$ must be symmetric as it is the Hessian of $f(\mathbf{h})$ and

$$\partial^2 f(\mathbf{h})/\partial h_i \partial h_j = \partial^2 f(\mathbf{h})/\partial h_j \partial h_i \qquad \text{for all paths } i \text{ and } j$$

Therefore $-\nabla \mathbf{d}(\mathbf{h})$ must also be symmetric if there is to be an equivalent optimisation problem. In general, of course, this is not the case.

For right angle cost functions, the variational inequality defining the optimum has as before the form

$$(\mathbf{h} - \mathbf{h}^*)^{\mathrm{T}}\mathbf{d} \geqslant 0 \qquad \text{for all feasible } \mathbf{h}$$

but the constraints are augmented by link capacities to become

$$\mathbf{t} = \mathbf{Bh}, \qquad \mathbf{h} \geqslant \mathbf{0}, \qquad \mathbf{s} \geqslant \mathbf{Ah}$$

The Jacobian of the path costs is indefinite because the link (and therefore the path) costs are unresponsive to changes in flow. However, the Jacobian of $\mathbf{d}(\mathbf{h})$ is still negative definite, provided the Jacobian of $\mathbf{g}^{\mathrm{D}}(\mathbf{h})$ is negative definite, or in other words, provided the demand for paths mainly falls with increasing cost. Hence, given this proviso, the equilibrium path flows $\mathbf{h}^*$ are still unique, *a very useful property in practice.*

6.3   A GENERAL EQUIVALENT OPTIMISATION PROBLEM

Under deterministic user equilibrium assignment, only minimum cost paths are used and paths that are not minimum cost are not used. Under stochastic user equilibrium assignment, however, only paths that are *perceived* to be minimum cost paths are chosen. It is not assumed that the perception of cost is perfect, rather that it varies randomly.

Sheffi (1985) has formulated a general equivalent optimisation problem for stochastic user equilibrium assignment, based on the expected minimum origin-destination costs, which is finding increasing use in the literature (see Maher and Hughes, 1995; Zhang and Smith, 1995). The objective function has the form

$$f(\mathbf{v}) = -\mathbf{t}^{\mathrm{T}} \mathbf{z}(\mathbf{v}) + \mathbf{v}^{\mathrm{T}} \mathbf{c}(\mathbf{v}) - \sum_{\text{all } i} \int_{x=0}^{x=v_i} c_i(x) \, dx$$

where $\mathbf{z}$ is the vector of expected minimum origin–destination costs and the summation is over all links. The objective function of course presupposes that the link cost functions are separable. If the relationship between link flows and link costs is also monotonic, the link cost functions may be inverted. Integration by parts yields

$$\sum_{i} \int_{x=c_{\min}}^{x=c_i} v_i(x) \, dx = \mathbf{v}^{\mathrm{T}} \mathbf{c}(\mathbf{v}) - \sum_{\text{all } i} \int_{x=0}^{x=v_i} c_i(x) \, dx$$

When the link flows are represented as functions of link costs, the above objective function therefore reduces to

$$f(\mathbf{c}) = -\mathbf{t}^{\mathrm{T}} \mathbf{z}(\mathbf{c}) + \sum_{\text{all } i} \int_{x=c_{\min}}^{x=c_i} v_i(x) \, dx$$

The gradient of the objective function with respect to link costs is

$$\nabla f(\mathbf{c}) = -\mathbf{t}^{\mathrm{T}} (\partial \mathbf{z}/\partial \mathbf{c}) + \mathbf{v}^{\mathrm{T}}$$

Note that the Jacobian of the expected minimum origin–destination costs with respect to link costs is equal to the matrix of link choice proportions, namely

$$\partial \mathbf{z}/\partial \mathbf{c} = \mathbf{P}^{\mathrm{T}}$$

because the effect of a *small* change of a link cost on an expected minimum origin–destination cost depends on the *probability* of a unit of flow (a trip) between the specified origin and destination using the link in question.

The Hessian of the objective function is

$$\nabla^2 f(\mathbf{c}) = \sum_{\text{all } w} -t_w (\partial z_w/\partial \mathbf{c} \partial \mathbf{c}) + \mathbf{J}^{-1}$$

where

$$\mathbf{J} = \partial \mathbf{c}/\partial \mathbf{v}$$

The Jacobian of the link cost functions is positive definite if the link cost functions are monotonically increasing, in which case the Jacobian is also invertible.

The *rate of change* of increase in expected minimum origin–destination costs with respect to increases in link costs will be zero or negative, as the probability of link choice decreases with increasing link cost. Hence the Hessian of the expected cost for origin–destination pair w with respect to link costs, namely

$$\partial z_w/\partial \mathbf{c} \partial \mathbf{c}$$

is negative semi-definite.

The Hessian of the objective function is therefore positive definite, since the sum of a series of positive semi-definite matrices and one positive matrix is itself positive definite. This in turn implies that the objective function is convex with a unique optimum where

$$\mathbf{t}^{\mathrm{T}}(\partial \mathbf{z}/\partial \mathbf{c}) = \mathbf{t}^{\mathrm{T}}\mathbf{P} = \mathbf{v}^{\mathrm{T}}$$

or when transposed

$$\mathbf{v} = \mathbf{Pt}$$

The optimum of $f(\mathbf{v})$, or alternatively $f(\mathbf{c})$, therefore defines a stochastic user equilibrium assignment. Note that the matrix of link choice proportions is a function of link cost which in turn is a function of link flow. The usefulness of this result lies principally in the gradients the objective function provides. Note that the assignment $\mathbf{Pt}$ resulting from the costs generated by the current link flows $\mathbf{v}$ always generates a decent direction, since

$$\nabla f(\mathbf{v})(\mathbf{Pt} - \mathbf{v}) = (\mathbf{v} - \mathbf{Pt})^{\mathrm{T}}\mathbf{J}(\mathbf{Pt} - \mathbf{v}) < 0 \qquad \text{when } \mathbf{v} \neq \mathbf{Pt}$$

This point is returned to in the case of the logit model.

6.4 LOGIT ASSIGNMENT

6.4.1 The logit path choice model

The logit path choice model has the form

$$\ln(h_j{}^*/h_{j'}{}^*) = -\alpha(g_j{}^* - g_{j'}{}^*)$$

where α is the *dispersion parameter*, and j and j' are two paths connecting the same origin–destination pair. The log of the ratio of the path flows is proportional to the difference in the path costs. This relationship applies to any pair of paths connecting a given origin and destination. Paths with equal costs get equal shares of traffic and paths with higher costs get lower shares than paths with lower costs. Note that for finite cost differences every path must have a positive share of traffic, reflecting the fact that under stochastic user equilibrium, as opposed to deterministic user equilibrium, *every path will be used*.

The logit model may be restated as follows:

$$\ln(h_j{}^*) + \alpha g_j{}^* = z_k$$

where path j connects origin–destination pair k and z_k is a factor specific to that origin–destination pair. Hence

$$\ln(\mathbf{h}^*) + \alpha \mathbf{g}(\mathbf{h}^*) = \mathbf{B}^{\mathrm{T}}\mathbf{z}$$

where $\mathbf{B}$ is the origin–destination–path incidence matrix.

If there is a function $f(\mathbf{h})$ for which

$$\nabla f(\mathbf{h}) = \ln(\mathbf{h}) + \alpha \mathbf{g}(\mathbf{h})$$

then

$$\nabla^2 f(\mathbf{h}) = [1/h_j] + \alpha \mathbf{A}^{\mathrm{T}}\mathbf{JA}$$

where $[1 / h_j]$ is a diagonal matrix with $1 / h_j$ in the jth position of the principal diagonal. The Hessian of $f(\mathbf{h})$ is thus positive definite provided the link costs are increasing in link flows, since the sum of a positive definite and a positive semi-definite matrix is a positive definite matrix. This in turn implies that $f(\mathbf{h})$ is convex.

The following optimisation problem

$$\text{Minimise } f(\mathbf{h}) \text{ subject to } \mathbf{t} = \mathbf{Bh}$$

is solved when

$$\nabla f(\mathbf{h}^*) = \mathbf{B}^\mathsf{T}\mathbf{z}$$

(see Chapter 3 on the optimality conditions for equality constrained optimisation problems), yielding the logit model.

A useful result is that at any stage of an iterative procedure for finding a logit assignment by solving the above minimisation problem, the logit assignment produced by the current vector of path flows constitutes a *descent direction* with respect to $f(\mathbf{h})$. To demonstrate this, let $\mathbf{h}'$ be the vector of current path flows. At this point, the gradient of the objective function is

$$\nabla f(\mathbf{h}') = \ln(\mathbf{h}') + \alpha\mathbf{g}(\mathbf{h}')$$

and the logit model yields

$$\ln(\mathbf{h}'') = \mathbf{B}^\mathsf{T}\mathbf{z} - \alpha\mathbf{g}(\mathbf{h}')$$

where the dual variables $\mathbf{z}$ are calculated so that

$$\mathbf{t} = \mathbf{Bh}$$

The gradient may be re-expressed as

$$\nabla f(\mathbf{h}') = \ln(\mathbf{h}') - \ln(\mathbf{h}'') + \mathbf{B}^\mathsf{T}\mathbf{z}$$

By pre-multiplication with the direction provided by the logit model

$$(\mathbf{h}'' - \mathbf{h}')^\mathsf{T}\nabla f(\mathbf{h}') = (\mathbf{h}'' - \mathbf{h}')^\mathsf{T}(\ln(\mathbf{h}') - \ln(\mathbf{h}'') + \mathbf{B}^\mathsf{T}\mathbf{z}) < 0$$

Since

$$(\mathbf{h}'' - \mathbf{h}')^\mathsf{T}\mathbf{B}^\mathsf{T} = \mathbf{0}^\mathsf{T}$$

then

$$(\mathbf{h}'' - \mathbf{h}')^\mathsf{T}\nabla f(\mathbf{h}') < 0$$

because the natural log transformation is concave, thereby demonstrating that $(\mathbf{h}'' - \mathbf{h}')$ is a descent direction. Thus the logit model combined with either a line search method or the method of successive averages can be used iteratively for finding the equilibrium logit assignment.

There is a combinatorial justification for the logit model. The logarithm of the number of possible *permutations* of trips giving rise to a given vector of path flows is

$$\ln(\Pi_{\text{all } w}t_w!/\Pi_{\text{all } j}h_j!) = \ln(\Pi_{\text{all } w}t_w!) - \ln(\Pi_{\text{all } j}h_j!) = \text{ constant } - \sum_{\text{all } j}\ln(h_j!)$$

The number of permutations that gives rise to $\mathbf{h}$ is referred to as the *entropy* of $\mathbf{h}$. As a result of Sterling's approximation it may be shown that $-\mathbf{h}^{\mathrm{T}}\ln(\mathbf{h})$ is approximately proportional to $-\sum_{\mathrm{all}\ j}\ln(h_j!)$. Thus the minimisation of $\mathbf{h}^{\mathrm{T}}\ln(\mathbf{h})$, or more conveniently of $\mathbf{h}^{\mathrm{T}}(\ln(\mathbf{h})-1)$, where $\mathbf{1}$ is a unit vector of appropriate dimension, is equivalent to maximising entropy. Note the $\mathbf{h}^{\mathrm{T}}\mathbf{1} = \mathbf{t}^{\mathrm{T}}\mathbf{1}$ is equal to the total number of trips in the trip table and is therefore fixed.

Consider the following convex optimisation problem

$$\text{Minimise } f(\mathbf{h}) = \mathbf{h}^{\mathrm{T}}(\ln(\mathbf{h})-1) \qquad \text{subject to } \mathbf{t} = \mathbf{Bh}, \quad C = \mathbf{c}^{\mathrm{T}}\mathbf{Ah}$$

where the link costs $\mathbf{c}$ are fixed and a total system cost constraint C is imposed. Both the objective function and constraints are convex, so there is a unique minimum. At the minimum

$$\nabla f(\mathbf{h}^*) = \ln(\mathbf{h}^*) = \mathbf{B}^{\mathrm{T}}\mathbf{z} + \mathbf{A}^{\mathrm{T}}\mathbf{c}\alpha = \mathbf{B}^{\mathrm{T}}\mathbf{z} + \mathbf{g}\alpha$$

where $\mathbf{z}$ are the dual variables relating to the origin–destination constraints, $\mathbf{g}$ is the vector of path costs and α is the dual variable relating to the total system cost constraint (see Chapter 3 on the optimality conditions for a convex programing problem). This implies that

$$\ln(\mathbf{h}^*) - \mathbf{g}\alpha = \mathbf{B}^{\mathrm{T}}\mathbf{z}$$

which as said earlier constitutes the logit model (the change of sign attached to α is not material).

The combinatorial justification for the logit path choice model is in fact rather weak, because it presupposes that each permutation of trips is equally likely. The value of the model as a predictive tool far outweighs the strength of this justification. Nonetheless, the equivalent optimisation problem that the entropy maximising problem gives rise to, turns out to be very useful in practice.

Logit choice models are frequently justified on the basis of *utility maximisation*, where perceived utility contains a random component, perhaps reflecting misperceptions or variations in taste (see Ben-Akiva and Lerman, 1985). Suppose that the perceived utility of a path j is given by the negative of its cost $-g_j$ and a random term ε_j as follows

$$u_j = -g_j + \varepsilon_j$$

then the probability that path j is chosen from a set of paths J is equal to

$$\Pr(u_j > u_k \text{ for all paths } k \text{ not equal to } j \text{ but in set } J)$$

When the random terms ε_j are identically and independently distributed with Gumbel distributions, then

$$\Pr(u_j > u_k \text{ for all paths } k \text{ not equal to } j \text{ but in set } J) = \exp(-\alpha g_j)/\sum_{k \in J}\exp(-\alpha g_k)$$

No convincing justification has been advanced for the assumption of Gumbel distributed utilities, other than that it yields a particularly tractable model, namely the logit model.

In the context of path choice, there is an objection to assuming independently distributed utilities, because alternative paths very often share links (see Daganzo and Sheffi, 1977). If the path choice is based on randomly distributed link utilities, then paths that share links will have correlated utilities. It may, however, be that the source of random variation is something else, for example the value of time.

One consequence of the random utility interpretation is that the expected utility of a set of paths $P(w)$, referred to by Sheffi (1985) as the *satisfaction* of $P(w)$, is given by the expected value of the maximum utility, which in the case of the logit model has the following explicit form:

$$E\{\text{Max}_{k \in P(w)} u_k\} = (1/\alpha) \ln \left(\sum_{k \in P(w)} \exp(-\alpha g_k) \right)$$

where $E\{.\}$ is the expected value of $\{.\}$ (see Williams, 1977). The expected minimum cost for origin–destination pair w is therefore

$$z_w = -E\{\text{Max}_{k \in P(w)} u_k\} = -(1/\alpha) \ln \left(\sum_{k \in P(w)} \exp(-\alpha g_k) \right)$$

where $P(w)$ is the set of paths connecting origin–destination pair w. This is frequently referred to as the *logsum* term, and plays an important role in the *hierarchical logit model*. If mis-perception occurs at the link level, one way to allow for the correlation between path utilities would be to make use of the hierarchical logit model. Alternative, paths with shared links can be combined by using the logsum term.

Note that the expected cost of travel from node m to node n is given by

$$z_{mn} = -(1/\alpha) \ln(v_{mn})$$

where v_{mn} is the element in row m and column n of the matrix

$$\mathbf{N} = \mathbf{W} + \mathbf{W}^2 + \mathbf{W}^3 + \ldots = (\mathbf{I} - \mathbf{W})^{-1} - \mathbf{I}$$

and matrix $\mathbf{W}$ has elements

$$w_{mn} = \begin{cases} \exp(-\alpha c_{mn}) & \text{if a link connects node } m \text{ to node } n \\ 0 & \text{otherwise} \end{cases}$$

6.4.2 Logit assignment with increasing cost functions

Two approaches are distinguished here, the first without the retention of path information and the second with the retention of path information. Algorithms not requiring the retention of path information are dealt with first.

When the cost functions are separable and monotonically non-decreasing in the link flows, then the objective function

$$f(\mathbf{h}) = E(\mathbf{h}) + \alpha B(\mathbf{v} = \mathbf{Ah})$$

with

$$E(\mathbf{h}) = \mathbf{h}^T (\ln(\mathbf{h}) - 1)$$

and

$$B(\mathbf{v}) = \sum_{i=1 \text{ to } N} \int_{x=0}^{x=v_i} c_i(x) \, dx$$

is strictly convex in $\mathbf{h}$, because the Hessian of $f(\mathbf{h})$ is positive definite. The convex optimisation problem

$$\text{Minimise } f(\mathbf{h}) \text{ subject to } \mathbf{t} = \mathbf{B}\mathbf{h}, \mathbf{h} \geqslant 0$$

generates the required gradient vector

$$\nabla f(\mathbf{h}) = \ln(\mathbf{h}) + \alpha \mathbf{g}(\mathbf{h})$$

For practical purposes, the non-negativity conditions can be neglected as the objective function is not defined for non-positive path flows. This optimisation problem was originally proposed by Fisk (1980).

The first practical solution method, proposed by Powell and Sheffi (1982), made use of the method of successive averages. Their algorithm has the following three steps:

Powell—Sheffi algorithm for logit assignment

Step 1 (initialisation)
  $\mathbf{v} \leftarrow \mathbf{0}$
  $n \leftarrow 1$

Step 2 (find a logit assignment for fixed link costs)
  $\mathbf{c} \leftarrow \mathbf{c}(\mathbf{v})$
  $\mathbf{v}^* \leftarrow$ logit assignment for $\mathbf{c}$

Step 3 (method of successive averages)
  $\mathbf{v} \leftarrow \mathbf{v}(1 - 1/n) + \mathbf{v}^*(1/n)$
  $n \leftarrow n + 1$
  if convergence insufficient then
      return to Step 2
  else stop.

The algorithm, at least as formulated here, starts with an empty network (Step 1). Based on the current link flows (initially zero), the current link costs are obtained. A fixed cost logit assignment yields the auxiliary link flows (Step 2). Powell and Sheffi (1982) propose the use of Dial's algorithm, set out in Dial (1971), but the alternative matrix inversion algorithm set out in Chapter 2 could also be used. The current and auxiliary link flows are then combined by the method of successive averages to obtain updated link flows (Step 3). If convergence is insufficient, the algorithm returns to Step 2 and repeats the process taking the updated link flows as the current link flows.

It has been shown in Section 6.2 that the logit model provides a descent direction with respect to Fisk's optimisation problem. Powell and Sheffi (1982) make use of Blum's Theorem to prove convergence of the procedure (presupposing of course that there is a solution to converge to).

A line search in Step 3 may improve the efficiency of the algorithm. The difficulty is that the first part of the objective function $f(\mathbf{h})$, namely $\mathbf{h}^{\mathrm{T}}(\ln(\mathbf{h}) - \mathbf{1})$, requires the definition of paths and the Powell–Sheffi algorithm does not retain paths. However, Akamatsu (1996) has shown that it is possible to reformulate the first part of $f(\mathbf{h})$ in terms of link flows disaggregated by origin *when no paths are excluded from the path set*. Note that if there are any cycles in the network, there will be infinitely many paths and the vector $\mathbf{h}$ will have infinitely many elements.

Under logit assignment, Akamatsu (1996) has noted that the probability of using any particular path can be expressed as the product of a series of conditional probabilities:

$$\text{Prob(choose path } r) = p(o|i)p(i|j)\ldots p(k|d)$$

where path r connects origin o with destination d via links $i, j, \ldots, k$ and where $p(i|j)$ is the conditional probability of coming from link i having arrived at link j. This is a *Markov chain*. Akamatsu (1996) demonstrates that as a result of the Markov property, maximising entropy across paths is equivalent to maximising the entropy *into* each link *for each origin centroid*. Thus in the logit assignment model the following

$$\left(\sum_{\text{all } l \text{ into } j} v_{ol}' \right)! / \Pi_{\text{all } l \text{ into } j} v_{ol}'!$$

is maximised for each link j and for each origin centroid o, where v_{ol}' is the flow on link l from origin centroid o. The summation and product is over all links l leading directly into link i. This can be shown, by Sterling's Approximation, to be equivalent to minimising

$$E(\mathbf{v}') = \sum_{\text{all } o} \left(\mathbf{v}_o'^T \ln(\mathbf{v}_o') - \sum_{\text{all } j} \left(\sum_{\text{all } l \text{ into } j} v_{ol}' \right) \ln \left(\sum_{\text{all } l \text{ into } j} v_{ol}' \right) \right)$$

where $\mathbf{v}'$ is now the vector of link flows $\mathbf{v}$ *disaggregated by origin centroid*. Furthermore,

$$\mathbf{v}'^T = [\mathbf{v}_1'^T, \mathbf{v}_2'^T, \ldots, \mathbf{v}_o'^T, \ldots]$$

where $\mathbf{v}_o'$ is that section of $\mathbf{v}'$ that specifically relates to origin centroid o.

Let

$$f(\mathbf{v}') = E(\mathbf{v}') + \alpha B(\mathbf{v}')$$

where $B(\mathbf{v}')$ is as defined before (note that link flows need to reaggregated before the evaluation of the link costs). Then

$$\partial f(\mathbf{v}')/\partial v_{oi}' = \ln \left(v_{oi}' / \sum_{\text{all } l \text{ into } i} v_{oi}' \right) + \alpha c_i(\mathbf{v})$$

where the summation is over all links l leading into link i. Thus

$$\nabla f(\mathbf{v}') = [r_{11} + \alpha c_1, r_{12} + \alpha c_2, \ldots, r_{21} + \alpha c_1, \ldots]^T$$

where for origin o and link i

$$r_{oi} = \ln \left(v_{oi}' / \sum_{\text{all } l \text{ into } i} v_{oi}' \right)$$

With the above definition of the gradients, the Frank–Wolfe algorithm presented earlier may be adapted as follows (this is the algorithm presented in Akamatsu, 1996, extended to encompass non-constant costs):

Frank–Wolfe algorithm for logit assignment without path retention

Step 1   (initialisation)

$\mathbf{v}''\leftarrow$ feasible non-zero disaggregated link flows

Step 2   (search direction)

$\mathbf{v}^*\leftarrow$ Minimise $LP(\mathbf{v}')=\mathbf{v}'^T\nabla f(\mathbf{v}'')$ subject to $\mathbf{v}_o=\mathbf{Ah}_o$ for all o,

$\mathbf{t}=\mathbf{Bh}$, $\mathbf{h}\geqslant 0$

Step 3   (line search)

$\mathbf{v}'\leftarrow$ Minimise $f(\mathbf{v}=\mathbf{v}'\lambda+\mathbf{v}^*(1-\lambda))$ subject to $1\geqslant\lambda\geqslant 0$

if convergence insufficient then

return to Step 2

else stop.

Note that $\mathbf{h}_o$ is $\mathbf{h}$ with all path flows *not* having origin o set to zero.

In comparison to the deterministic user equilibrium case, there are many more variables and the link costs have been augmented by a negative quantity specific to each origin centroid. To avoid problems with infinite cycles, the dispersion parameter may have to be increased (meaning less dispersion) to keep the elements of $\nabla f(\mathbf{v}'')$ non-negative. As before, an all-or-nothing assignment may be used to solve Step 2 without having to retain any path information. The difference here is that the costs used to generate the least cost trees are specific to each origin.

As argued earlier, traffic engineers and transportation planners are frequently interested in path flows, so an algorithm which retains this information is in principle attractive. It may also be more efficient. The generation and retention of path information is referred to as *column generation*, because columns of matrices $\mathbf{A}$ and $\mathbf{B}$ are generated as the iterations progress.

There are many approaches to column generation, depending of the features of the paths required. Clearly there should be sufficient paths generated to cater for all the trips specified by the trip table. The paths generated should be the more significant paths in terms of their flows. The generation of all paths is generally not practical and, in the presence of cycles, not possible.

Bell *et al.* (1993) proposed a Frank–Wolfe algorithm with augmented link costs, but, unlike Akamatsu (1996), did not require the disaggregation of link flows by origin. Instead the following *auxiliary problem*

$$\text{Minimise } E(\mathbf{h})=\mathbf{h}^T(\ln(\mathbf{h})-1)\text{ subject to }\mathbf{v}=\mathbf{Ah}$$

is solved for the vector of current link flows, $\mathbf{v}$. The dual variables of this problem give the gradient of the minimum of $\mathbf{h}^T(\ln(\mathbf{h})-1)$ with respect to $\mathbf{v}$. Note that the solution of the auxiliary problem is given by

$$\ln(\mathbf{h}^*)=\mathbf{A}^T\mathbf{u}$$

or equivalently

$$\mathbf{h}^*=\exp(\mathbf{A}^T\mathbf{u})$$

The dual variables are obtained by solving the simultaneous equations

$$\mathbf{v}=\mathbf{A}\,\exp(\mathbf{A}^T\mathbf{u}^*)$$

for $\mathbf{u}^*$, which may be done by iterative balancing. If there is a feasible solution, convergence is assured (see Chapter 3 for a description of the iterative balancing method and for a general proof of convergence). The algorithm has the following form:

Frank—Wolfe algorithm for logit assignment with path retention

Step 1 (initialisation)
        $\mathbf{v} \leftarrow$ All-or-nothing assignment on $\mathbf{c}(\mathbf{0})$
        $\mathbf{A}, \mathbf{B} \leftarrow$ Paths

Step 2 (search direction)
        $\mathbf{u}^* \leftarrow$ Minimise $\mathbf{h}^T(\ln(\mathbf{h}) - 1)$ subject to $\mathbf{v} = \mathbf{Ah}$
        $\mathbf{c}^* \leftarrow \mathbf{c}(\mathbf{v}) + (1/\alpha)\mathbf{u}^*$
        $\mathbf{v}^* \leftarrow$ All-or-nothing assignment on $\mathbf{c}^*$
        $\mathbf{A}, \mathbf{B} \leftarrow \mathbf{A}, \mathbf{B} +$ New paths

Step 3 (line search)
        $\mathbf{v} \leftarrow$ Minimise $f(\mathbf{v}^*\lambda + \mathbf{v}(1 - \lambda))$ subject to $1 \geqslant \lambda \geqslant 0$
        if convergence insufficient then
            return to Step 2
        else stop.

As a result of the auxiliary optimisation problem, this algorithm is more complex than preceding ones. In Step 1, starting with the empty network link costs, an all-or-nothing assignment yields the current link flows. The paths generated by the all-or-nothing assignment constitute the initial columns of incidence matrices $\mathbf{A}$ and $\mathbf{B}$. In Step 2, the auxiliary problem yields the gradient of maximum entropy (or rather that of the negative of its logarithm) for the current link flows. The augmented link costs are formed. An all-or-nothing assignment yields the auxiliary link flows. The new paths are added to matrices $\mathbf{A}$ and $\mathbf{B}$. In Step 3, a line search yields new flows. Function evaluations in the line search involve solving the auxiliary problem. Alternatively, the method of successive averages may be used. If convergency is insufficient, the algorithm returns to Step 2, taking the new link flows as the current link flows.

Concerning the interpretation of the model, the logit assignment aims to achieve equality of $\ln(h_j) + \alpha g_j$ across the paths j connecting one trip-table element. Hence, $(1/\alpha) \ln(h_j)$ can be interpreted as the mean value of the perceived error of path j's cost. Likewise, the second term of the augmented link costs, $(1/\alpha)\mathbf{u}^*$, is the mean value of the perceived error of the links' costs.

The numerical example presented in Chen and Alfa (1991) is adopted here to illustrate the augmented link cost approach. The test network is shown in Fig. 6.4, where the numbers against the links are the corresponding free flow travel costs. It consists of 17 links and 12 nodes. There is only one origin–destination pair and the travel demand is 10 units (say vehicles per minute). The following link cost function was adopted for all links:

$$c_i(x_i) = t_i + 0.008v_i^4$$

where t_i is the free flow travel cost on link i. Five α values, ranging from 0.01 to 10.0, have been chosen for testing.

The computational results in terms of the objective function value, path flow entropy $E(\mathbf{h})$, integral network cost $B(\mathbf{v})$, average travel cost, number of iterations required and number of paths generated are shown in Table 6.1 with respect to the various values of α. It can be seen that the greater the α value, the closer the objective function value

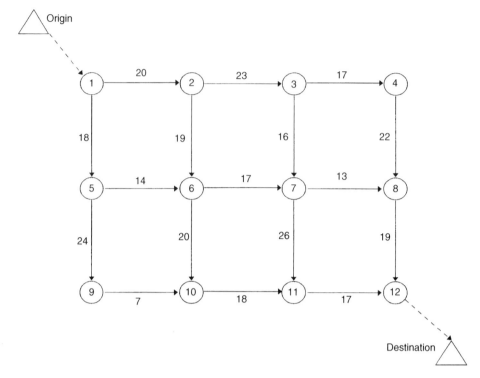

Figure 6.4 *Example network*

Table 6.1 *Computational results for augmented link cost algorithm*

Parameter α	Objective function value $E(h)/\alpha + B(v)$	Path flow entropy $E(h)$	Integral network cost $B(v)$	Average travel cost	Number iterations to converge	Number of of paths
0.01	884.7	−0.3687	921.4	109.7	6	4
0.05	904.3	−0.3674	911.6	109.6	7	3
0.1	904.4	−0.3227	907.6	106.4	8	4
1.0	899.2	−0.3416	899.6	104.2	20	5
10.0	900.2	−0.2905	900.2	104.6	16	5

is to the integral network cost. However, the path flow entropy, integral network cost and average travel cost may not be monotonic in α as different sets of paths could be generated by the proposed algorithm for different α. It can be shown that the path flow entropy and the integral network cost are monotonically decreasing with respect to α *if the same set of paths is used*, providing a basis for the estimation of α.

6.4.3 Logit assignment with right angle cost functions

When right angle cost functions are used, a logit assignment is found by solving the following form of Fisk's optimisation problem

$$\text{Minimise } f(\mathbf{h}) = \mathbf{h}^{\mathrm{T}}(\ln(\mathbf{h}) - 1) + \alpha \mathbf{c}^{\mathrm{T}} \mathbf{A} \mathbf{h} \text{ subject to } \mathbf{t} = \mathbf{B}\mathbf{h}, \ \mathbf{s} \geqslant \mathbf{A}\mathbf{h}, \ \mathbf{h} \geqslant \mathbf{0}$$

where, as before, **s** are link capacities. For practical purposes the non-negativity conditions may be neglected. Both the objective function and the constraints are convex, so there is a unique solution to the problem in terms of the primal variables, **h**.

The corresponding Lagrangian equation has the form

$$L(\mathbf{h}, \mathbf{u}, \mathbf{z}) = \mathbf{h}^{\mathrm{T}}(\ln(\mathbf{h}) - 1) + \alpha \mathbf{c}^{\mathrm{T}}\mathbf{Ah} - \mathbf{u}^{\mathrm{T}}(\mathbf{s} - \mathbf{Ah}) + \mathbf{z}^{\mathrm{T}}(\mathbf{t} - \mathbf{Bh})$$

and the optimality conditions are

$$\ln(\mathbf{h}) + \mathbf{A}^{\mathrm{T}}\mathbf{c}\alpha + \mathbf{A}^{\mathrm{T}}\mathbf{u} - \mathbf{B}^{\mathrm{T}}\mathbf{z} = 0$$

$$\mathbf{s} - \mathbf{Ah} \geqslant 0$$

$$\mathbf{u} \geqslant 0$$

$$\mathbf{u}^{\mathrm{T}}(\mathbf{s} - \mathbf{Ah}) = 0$$

$$\mathbf{t} - \mathbf{Bh} = 0$$

The first and last of these conditions provide the logit model, because after some rearrangement,

$$h_j = t_w \exp(-\alpha g_j - \mathbf{a}_j^{\mathrm{T}}\mathbf{u}) / \sum_j \exp(-\alpha g_j - \mathbf{a}_j^{\mathrm{T}}\mathbf{u})$$

where $\mathbf{a}_j$ is the jth column of the link–path incidence matrix **A**. As the elements of **u** are only positive when the flow of the corresponding link has reached capacity, it would seem intuitively reasonable to associate them with delay. In fact it can be shown that the elements of **u** may be associated with equilibrium link delays divided by α, the dispersion parameter. The elements of **z**, however, no longer correspond to the costs of the least cost paths between each origin and destination, as in the case of deterministic user equilibrium assignment, but are proportional to the origin–destination demand.

The interpretation of the dual variables presented here makes use of two facts; if a small quantity of flow is subtracted from a path there can be no active capacity constraint on that path, and no path can have more than one active capacity constraint before the quantity of flow is subtracted. Consider a path j with an active capacity constraint on link i. A small relaxation of this constraint causes a small quantity of flow to move to path j from an alternative path. The small quantity of flow would come directly or indirectly from a path with no active constraint *ex post*, say path j'. For path j, the objective function changes by

$$\Delta f(\mathbf{h}) = (h_j + \delta)(\ln(h_j + \delta) - 1) - h_j(\ln(h_j) - 1) + \alpha\delta g_j$$

where δ represents a small positive amount of flow. By a first order (linear) approximation

$$(h_j + \delta)(\ln(h_j + \delta) - 1) - h_j(\ln(h_j) - 1) \approx \delta \ln(h_j)$$

so

$$(h_j + \delta)(\ln(h_j + \delta) - 1) - h_j(\ln(h_j) - 1) + \alpha\delta g_j \approx \delta \ln(h_j) + \alpha\delta g_j$$

For path j', the objective function changes by

$$(h_{j'} - \delta)\ln(h_{j'} - \delta) - 1) - h_{j'}(\ln(h_{j'}) - 1) - \alpha\delta g_{j'} \approx -\delta \ln(h_{j'}) - \alpha\delta g_{j'}$$

The dual variable of the constraint indicates the rate at which the optimum value of the objective function reduces as the constraint is relaxed, so

$$\delta \ln(h_j) + \alpha \delta g_j - \delta \ln(h_{j'}) - \alpha \delta g_{j'} \approx -\delta u_i$$

As δ tends to zero,

$$\ln(h_j/h_{j'}) = -\alpha(g_j - g_{j'}) - u_i$$

Since path j has one bottleneck at link i, and path j' has no bottleneck after the transfer of δ trips across to path j, there will be a delay on path j, which in turn must be the delay on link i. When this delay is u_i/α, there is a logit assignment. Hence the elements of $\mathbf{u}$ are interpretable as link delays divided by α.

As indicated in Chapter 3, the elements of $\mathbf{u}$ are only unique if the corresponding constraints are linearly independent, which in turn implies that only linearly independent links should be included. The number of linearly independent links is equal to the total number of links minus the number of internal nodes, as noted earlier.

The interpretation of the elements of $\mathbf{z}$ is as follows. From the optimality conditions

$$\mathbf{B}^T \mathbf{z}^* = \ln(\mathbf{h}^*) + \mathbf{g}\alpha + \mathbf{A}^T \mathbf{u}^* = \ln(\mathbf{h}^*) + \alpha \mathbf{g}'$$

where $\mathbf{g}'$ is the vector of equilibrium path costs inclusive of delay. For any particular path j and corresponding origin–destination pair w,

$$h_j^* = \exp(z_w^* - \alpha g_j')$$

Thus

$$t_w = \exp(z_w^*) \sum_{j \in P(w)} \exp(-\alpha g_j')$$

and so

$$z_w^* = -\ln\left(\sum_{j \in P(w)} \exp(-\alpha g_j')\right) + \ln(t_w)$$

Hence z_w^* is related to the expected minimum origin–destination cost in the following way:

$$z_w^* = \alpha \text{ (expected minimum origin–destination cost)} + \ln(t_w)$$

The dual variables may be estimated by iterative balancing, described in Chapter 3, provided the link–path and path–origin–destination matrices, $\mathbf{A}$ and $\mathbf{B}$, are available. The algorithm presented in Bell (1995a) finds the dual variables by iterative balancing.

Iterative balancing algorithm for logit assignment

Step 1 (initialisation)
$\quad\quad \mathbf{u} \leftarrow \mathbf{0}$
$\quad\quad \mathbf{z} \leftarrow \mathbf{0}$

Step 2 (iterative balancing)
$\quad\quad$ repeat
$\quad\quad\quad\quad$ for all origin—destination pairs w
$\quad\quad\quad\quad\quad\quad \ln(\mathbf{h}) \leftarrow -\mathbf{A}^T \mathbf{c}\alpha + \mathbf{B}^T \mathbf{z} - \mathbf{A}^T \mathbf{u}$
$\quad\quad\quad\quad\quad\quad$ if $t_w \neq b_w^T \mathbf{h}$ then $z_w \leftarrow z_w + \ln(t_w) - \ln(b_w^T \mathbf{h})$

```
for all links i
    ln(h) ← -AᵀCα + Bᵀz - Aᵀu
    if sᵢ < aᵢᵀ h then uᵢ ← uᵢ - ln(sᵢ) + ln(aᵢᵀ h)
until convergence.
```

Note that $\mathbf{a}_i^T$ is the ith row of matrix $\mathbf{A}$ and that $\mathbf{b}_w^T$ is the wth row of matrix $\mathbf{B}$. Convergence may be assessed in terms of relative changes in primal or dual variables or degree of constraint compliance. The proof of convergence given in Chapter 3 may be adapted, noting that $\mathbf{A}^T\mathbf{c}\alpha$ is a vector of constants.

The example presented in Chapter 5 is reconsidered. The network is shown in Fig. 5.5. There is one origin–destination pair, six links, four nodes and four paths. The costs and capacities of the links are shown in Table 6.2. The total demand from the origin to the destination is 4 units of flow (say vehicles per second). The link–path incidence matrix is

$$\mathbf{A} = \begin{bmatrix} 1 & 0 & 0 & 1 \\ 0 & 1 & 1 & 0 \\ 1 & 0 & 0 & 0 \\ 0 & 0 & 1 & 0 \\ 1 & 1 & 0 & 0 \\ 0 & 0 & 1 & 1 \end{bmatrix}$$

and the origin–destination–path matrix is

$$\mathbf{B} = \begin{bmatrix} 1 & 1 & 1 & 1 \end{bmatrix}$$

In Table 6.3, the link flows and delays are contrasted for three different values of α. The deterministic user equilibrium link flows are also given for comparison. As α is reduced, sensitivity to cost is reduced, so the delay required to hold demand at capacity on link 1 increases.

In Table 6.4, the path flows are compared with the path costs and the total system cost for three values of α. The demand for the most expensive path, path 3, falls with increasing α. Furthermore, the total system cost falls moderately with increasing α. Increasing sensitivity to cost does not of course imply the minimisation of total cost.

Where the incidence matrices are not already given, paths must be built (the columns of $\mathbf{A}$ and $\mathbf{B}$ must be generated). This is done by the use of a shortest path algorithm in some iterative way. It is not practical to generate all used paths, as *all* paths would be used under stochastic user equilibrium assignment and in a network with loops there would be infinitely many paths. The objective is therefore to generate the paths that are

Table 6.2 *Link costs and capacities*

Link	Cost	Capacity
1	1	1
2	3	4
3	1	2
4	3	4
5	1	2
6	3	4

Table 6.3 *Link flows and delays*

	$\alpha = 1$		$\alpha = 0.1$		$\alpha = 0.01$		
Link	Flow	Delay	Flow	Delay	Flow	Delay	DUE
1	1	5.0674	1	14.0862	1	112.8712	1
2	3	0	3	0	3	0	3
3	0.0337	0	0.4254	0	0.4925	0	1
4	1.0337	0	1.4254	0	1.4925	0	2
5	2	4.3569	2	4.0050	2	4.0000	2
6	2	0	2	0	2	0	2

Table 6.4 *Path flows and cost*

	Path flows			
Path	$\alpha = 1$	$\alpha = 0.1$	$\alpha = 0.01$	Cost
1	0.0337	0.4254	0.4925	3
2	1.9663	1.5746	1.5075	4
3	1.0337	1.4254	1.4925	9
4	0.9663	0.5746	0.5075	4
Total cost	21.1347	22.7017	22.9700	

most likely to be used. This may involve taking into account factors not included in the network specification. For example, drivers may be discouraged in various ways from using certain paths through residential areas.

In Bell (1995a), the following capacity-oriented column generation scheme is proposed. In essence, it builds and fills the least cost paths first, only building new paths when paths with less cost than those already existing can be found. Delay is included in the calculation of cost.

Capacity-oriented column generation for logit assignment

Step 1   (initialisation)
$$\mathbf{u} \leftarrow \mathbf{0}$$

Step 2   (generate columns)
        repeat
            $\mathbf{c}' \leftarrow \mathbf{c} + \mathbf{u}/\alpha$
            $\mathbf{A}, \mathbf{B} \leftarrow$ least cost paths with respect to $\mathbf{c}'$
            $\mathbf{u}, \mathbf{v}, \mathbf{h} \leftarrow$ iterative balancing
        until no new paths are built.

Initially the dual variables are initialised to zero (Step 1), thereby assuming that there are no delays on any links. Then least cost paths are built and dual variables determined (Step 2). The delay arising at bottlenecks may be sufficient to lead to the building of new least cost paths. In the early stages of Step 2, there may be insufficient capacity in the existing set of paths for the trip table to be loaded, in which case the iterative balancing algorithm cannot converge. The dual variables corresponding to bottleneck links will tend to infinity as the balancing progresses. In this case, iterative balancing is terminated and high delay costs are assigned to the bottleneck links. This forces the generation of new paths circumventing the bottlenecks, if such paths can be found. If no such paths can be found, there is no feasible solution. The algorithm terminates when no new least cost paths can be found.

The above column generation scheme is *conservative* in the sense that it terminates when no new *least* cost paths can be found. There may be other paths that would significantly increase entropy (reduce $\mathbf{h}^T(\ln(\mathbf{h}) - 1)$). The augmented link cost approach, developed for increasing link cost functions in the preceding section, may be applied here as well. This implies solving the following auxiliary problem:

$$\text{Minimise } E(\mathbf{h}) = \mathbf{h}^T(\ln(\mathbf{h}) - 1) \text{ subject to } \mathbf{v} = \mathbf{Ah}$$

for the current set of link flows. The dual variables for this problem $\mathbf{w}$ are then added to α times the current link costs, which include the current delays. This leads to the following revised column generation scheme, which would generate more paths.

Revised iterative column generation for logit assignment

Step 1   (initialisation)
        $\mathbf{u} \leftarrow \mathbf{0}$
        $\mathbf{w} \leftarrow \mathbf{0}$

Step 2   (generate columns)
        repeat
            $\mathbf{c}' \leftarrow \mathbf{w} + \mathbf{c}\alpha + \mathbf{u}$
            A, B $\leftarrow$ least cost paths with respect to $\mathbf{c}'$
            u, v, h $\leftarrow$ solve logit assignment problem by
            iterative balancing
            w $\leftarrow$ solve auxiliary problem by iterative balancing
        until no new paths are built.

6.5 PROBIT ASSIGNMENT

6.5.1 Probit model

For two paths connecting the same origin–destination pair, the probit model has the form

$$\Phi^{-1}(h_i{}^*/(h_i{}^* + h_j{}^*)) = -\alpha(g_i{}^* - g_j{}^*)$$

where $\Phi(x)$ is the integral of the unit normal distribution from 0 to x and $\Phi^{-1}(.)$ is the inverse of this function, referred to as the *probit function*. For statistically independent alternatives (namely, when the probability of choosing path i over path j is unrelated to the probability of choosing any third path k) the probit function behaves like the logit function. However, to neglect the correlations between choice probabilities is to ignore one of the major advantages of the probit approach.

The kind of network where the probit model would be attractive in relation to the logit model is demonstrated in Fig. 6.5. The choice probability for path 1 (links 2 and 3) is related to that for path 2 (links 2 and 4) because paths 1 and 2 share link 2. This relationship influences the probability of choosing path 3 (link 1). The choice process is probably better represented here by a nested structure, whereby the trip-maker first chooses between paths 1 or 2 and path 3. If the trip-maker opts for paths 1 or 2 he then chooses between them. The consequence of ignoring the nested choice structure is to overload the shared link (see Daganzo and Sheffi, 1977).

The difficulty with the probit model of path choice is that there is no known equivalent optimisation problem.

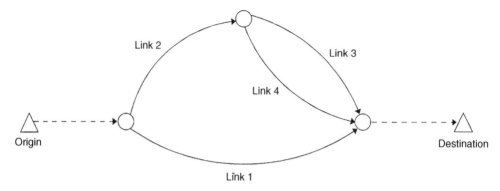

Figure 6.5   *Example network with correlated path choices*

6.5.2   *Probit assignment algorithms*

There are two approaches to probit assignment. The most commonly used approach is based on Monte Carlo simulation. Link costs are randomised in some appropriate way and then a deterministic user equilibrium assignment is sought. The probit assignment is found by averaging a sequence of deterministic user equilibrium assignments. The randomisation is a simulation of error in perceptions. Since link costs are randomised, correlations in perception between paths that share links are taken into account. Some argue that this is more realistic than the assumption of independence implicit in the logit assignment.

Monte Carlo simulation, while conceptually simple, will be computationally demanding. A reasonable number of random samples would be required in order to establish average values that are sufficiently independent of the random number generator. The Monte Carlo method has been used extensively by Daganzo (1982), Sheffi (1985) and others. Before its theoretical properties were fully understood, the method had been suggested by Burrell (1968) for the case of links with constant costs.

The alternative to Monte Carlo simulation is the use of Clarke's approximation. This gives the mean and variance for the *minimum* value of two variables that are approximately normally distributed with known means, variances and covariance. In the case of the network shown in Fig. 6.5, Clarke's approximation allows the probability of choosing link 1 over link 2 to be calculated because the mean and variance for the *minimum* of the perceived costs for links 3 and 4 can be added to the mean and variance of perceived cost for link 2 (note variances are additive for independent variables). Comparing the result with the mean and variance for the perceived cost of link 1 allows the probability of choosing link 1 over link 2 to be calculated (approximately). The use of Clarke's approximation is discussed in Sheffi (1985). More recently, Maher and Hughes (1995) have been applying the approximation in a tree-building algorithm to obtain a probit assignment.

6.6   *SENSITIVITY*

As there is no known equivalent optimisation problem for probit assignment, a sensitivity analysis with respect to perturbations in demand would require extensive Monte Carlo simulation. However, for the logit assignment model with increasing link cost functions, the sensitivity expression derived in Chapter 3 may be used. Noting that the Hessian of the objective function of the equivalent optimisation problem has the following form:

$$\nabla^2 f(\mathbf{h}^*) = \begin{bmatrix} 1/h_1{}^* & & \\ & 1/h_2{}^* & \\ & & \ddots \end{bmatrix} + \mathbf{A}^T \nabla \mathbf{c}(\mathbf{v}^*) \mathbf{A}$$

the effects of perturbations in the trip table on the equilibrium path flows are given by

$$\Delta \mathbf{h}^* = (\nabla^2 f(\mathbf{h}^*))^{-1} \mathbf{B}^T (\mathbf{B}(\nabla^2 f(\mathbf{h}^*))^{-1} \mathbf{B}^T)^{-1} \Delta \mathbf{t}$$

As the number of paths will equal or exceed the number of elements in the trip table, there should be no problem with matrix inversion.

Alternatively, the Tobin and Friesz (1988) approach to the analysis of sensitivities may be applied. At optimality

$$\ln(\mathbf{h}^*) + \alpha \mathbf{g}(\mathbf{h}^*) = \mathbf{B}^T \mathbf{z}^*$$

(the star denotes an equilibrium). The following relationship therefore holds for small changes of the primal and dual variables, $\Delta \mathbf{h}$ and $\Delta \mathbf{z}$, respectively,

$$\Delta \mathbf{g} \alpha = \mathbf{D} \Delta \mathbf{h} + \mathbf{B}^T \Delta \mathbf{z}$$

where $\mathbf{D} = -\nabla^2 f(\mathbf{h}^*)$. Additionally

$$\mathbf{t} = \mathbf{B} \mathbf{h}$$

so for small perturbations

$$\Delta \mathbf{t} = \mathbf{B} \Delta \mathbf{h}$$

Note that while $\mathbf{t}$ is related only to the primal variables, $\mathbf{g}$ is related to both the primal and dual variables. In vector and matrix notation

$$\begin{bmatrix} \Delta \mathbf{g} \alpha \\ \Delta \mathbf{t} \end{bmatrix} = \begin{bmatrix} \mathbf{D} & \mathbf{B}^T \\ \mathbf{B} & \mathbf{0} \end{bmatrix} \begin{bmatrix} \Delta \mathbf{h} \\ \Delta \mathbf{z} \end{bmatrix}$$

is the *first order* effect of perturbations in the primal and dual variables on the path costs and the trip table. The matrix between the two perturbation vectors is referred to as the *Jacobian of the system*. The inverse of this Jacobian provides the required sensitivities. Writing the inverse Jacobian in the corresponding block form

$$\begin{bmatrix} \Delta \mathbf{h} \\ \Delta \mathbf{z} \end{bmatrix} \begin{bmatrix} \mathbf{J}_{11} & \mathbf{J}_{12} \\ \mathbf{J}_{21} & \mathbf{J}_{22} \end{bmatrix} \begin{bmatrix} \Delta \mathbf{g} \alpha \\ \Delta \mathbf{t} \end{bmatrix}$$

it can be shown that

$$\mathbf{J}_{11} = \mathbf{D}^{-1}(\mathbf{I} - \mathbf{B}^T(\mathbf{B}\mathbf{D}^{-1}\mathbf{B}^T)^{-1}\mathbf{B}\mathbf{D}^{-1})$$

$$\mathbf{J}_{12} = \mathbf{D}^{-1}\mathbf{B}^T(\mathbf{B}\mathbf{D}^{-1}\mathbf{B}^T)^{-1}$$

$$\mathbf{J}_{21} = (\mathbf{B}\mathbf{D}^{-1}\mathbf{B}^T)^{-1}\mathbf{B}\mathbf{D}^{-1}$$

$$\mathbf{J}_{22} = -(\mathbf{B}\mathbf{D}^{-1}\mathbf{B}^T)^{-1}$$

Matrix $\alpha \mathbf{J}_{11}$ gives the sensitivity of path flows to perturbations in path costs and $\mathbf{J}_{12}$ gives the sensitivity of path flows to perturbations in the trip table.

In the case of right angle cost functions, the assignment yielded by the logit path choice model has been determined by the trip table $\mathbf{t}$, the active link capacity constraints $\mathbf{s}_a$ and the path costs $\mathbf{g}$. The sensitivity of the fitted path flows to these inputs may be analysed using an extension of the Tobin and Friesz (1988) method, provided all constraints are linearly independent.

At optimality the following relationships hold for small changes in the primal and dual variables, $\Delta \mathbf{h}$, $\Delta \mathbf{u}_a$ and $\Delta \mathbf{z}$ respectively,

$$\Delta \mathbf{g} \alpha = \mathbf{D} \Delta \mathbf{h} + \mathbf{B}^T \Delta \mathbf{z} - \mathbf{A}_a^T \Delta \mathbf{u}_a$$

$$\Delta \mathbf{t} = \mathbf{B} \Delta \mathbf{h}$$

$$\Delta \mathbf{s}_a = \mathbf{A}_a \Delta \mathbf{h}$$

where $\mathbf{D}$ is as before $-\nabla^2 f(\mathbf{h}^*)$. Subscript a identifies the set of links with active capacity constraints. Small changes in the path costs, the trip table elements and the active capacity constraints are related to small changes in the primal and dual variables as follows:

$$\begin{bmatrix} \Delta \mathbf{g} \\ \Delta \mathbf{t} \\ \Delta \mathbf{s}_a \end{bmatrix} = \begin{bmatrix} \mathbf{D}/\alpha & \mathbf{B}^T/\alpha & -\mathbf{A}_a^T/\alpha \\ \mathbf{B} & 0 & 0 \\ \mathbf{A}_a & 0 & 0 \end{bmatrix} \begin{bmatrix} \Delta \mathbf{h} \\ \Delta \mathbf{z} \\ \Delta \mathbf{u}_a \end{bmatrix}$$

The matrix between the two perturbation vectors is the Jacobian of the system. The inverse of this Jacobian provides the required sensitivities, *provided the constraints included are linearly independent*. The inverse of this Jacobian, represented in block form below, gives the sensitivities:

$$\begin{bmatrix} \Delta \mathbf{h} \\ \Delta \mathbf{z} \\ \Delta \mathbf{u}_a \end{bmatrix} = \begin{bmatrix} \mathbf{J}_{11} & \mathbf{J}_{12} & \mathbf{J}_{13} \\ \mathbf{J}_{21} & \mathbf{J}_{22} & \mathbf{J}_{23} \\ \mathbf{J}_{31} & \mathbf{J}_{32} & \mathbf{J}_{33} \end{bmatrix} \begin{bmatrix} \Delta \mathbf{g} \\ \Delta \mathbf{t} \\ \Delta \mathbf{s}_a \end{bmatrix}$$

Hence, assuming that the active capacity constraints remain active and the inactive capacity constraints remain inactive, the perturbations have the following effect on path flows:

$$\Delta \mathbf{h} = \mathbf{J}_{11} \Delta \mathbf{g} + \mathbf{J}_{12} \Delta \mathbf{t} + \mathbf{J}_{13} \Delta \mathbf{s}_a$$

Variances and covariances may be obtained approximately from the sensitivities. In particular, when path costs and link capacities remain unaltered,

$$\text{Var}\{\mathbf{h}^*\} \approx \mathbf{J}_{12} \text{Var}\{\mathbf{t}\} \mathbf{J}_{12}^T$$

As an example, consider the network in Fig. 7.1. The link capacities, costs, flows, dual variables and delays are given in Table 6.5 and the trip table, dual variables and shadow prices are given in Table 6.6. The paths and their fitted flows are given in Table 6.7. The value of α used was 0.1.

The sensitivity of path flows to the trip table are shown in Table 6.8. The effects of the capacity constraints on links 3 and 4 are evident. Note that, as one might expect, the sensitivities for each origin–destination pair sum to one, reflecting the conservation of flow (path flows must sum to the trip table).

Table 6.5 *Capacities, costs, flows and delays*

Link	Capacity	Cost	Flow	Dual	Delay
1	15	5	10	0	0
2	15	5	10	0	0
3	3	2	3	0.6978	6.9782
4	3	2	3	0.6978	6.9782
5	10	5	3.9816	0	0
6	10	5	3.9816	0	0

Table 6.6 *Trip table, duals and shadow prices*

O–D pair	Flow	Dual	Shadow price
A to C	5	1.8886	18.8859
A to D	5	2.0955	20.9555
B to C	5	2.0955	20.9555
B to D	5	1.8886	18.8859

Table 6.7 *Path flows*

Path	1	2	3	4	5	6	7	8
Links	1	2,3,6	2	1,4,5	2,3	1,5	1,4	2,6
Flow	4.0092	0.9908	4.0092	0.9908	2.0092	2.9908	2.0092	2.9908

Table 6.8 *Path flow sensitivity to the trip table*

	Origin-destination pair			
Path	A to C	A to D	B to C	B to D
1	0.881	0.160	0	0
2	0.119	−0.160	0	0
3	0	0	0.160	0.881
4	0	0	−0.160	0.119
5	−0.119	0.160	0	0
6	0.119	0.840	0	0
7	0	0	0.160	−0.119
8	0	0	0.840	0.119

Table 6.9 *Path flow sensitivity to path cost*

Path	1	2	3	4	5	6	7	8
1	−0.0478	0.0478	0	0	−0.0478	0.0478	0	0
2	0.0478	−0.0478	0	0	0.0478	−0.0478	0	0
3	0	0	−0.0478	0.0478	0	0	−0.0478	0.0478
4	0	0	0.0478	−0.0478	0	0	0.0478	−0.0478
5	−0.0478	0.0478	0	0	−0.0478	0.0478	0	0
6	0.0478	−0.0478	0	0	0.0478	−0.0478	0	0
7	0	0	−0.0478	0.0478	0	0	−0.0478	0.0478
8	0	0	0.0478	−0.0478	0	0	0.0478	−0.0478

Table 6.10 *Path flow sensitivity to the active capacity constraints*

Path	Active capacity constraint	
	Link 3	Link 4
1	−0.398	0
2	0.398	0
3	0	−0.398
4	0	0.398
5	0.602	0
6	−0.602	0
7	0	0.602
8	0	−0.602

The sensitivity of the path flows to path costs is shown in Table 6.9. The effects of trip table and the active capacity constraints on links 3 and 4 are reflected in the block structure of the sensitivity matrix. Note that the column sums are all zero, because of the origin–destination total constraints.

The sensitivity of fitted path flows to the active capacity constraints is shown in Table 6.10. As a result of the trip table constraints, the sensitivities sum to zero across all the paths.

6.7 ELASTIC DEMAND

The stochastic user equilibrium assignment model as presented so far has taken as input a fixed trip table. Concern has been shown recently about the use of fixed trip tables when assessing the traffic consequences of major infrastructure projects, because of the effect such projects can have on the generation and distribution of traffic. The new traffic resulting from such projects is often referred to collectively as *induced traffic*. One way of allowing for this phenomenon is to let the number of trips between each origin and destination be a function of the expected minimum cost of travel between the origin and the destination, on the basis that trip-makers will choose the path with the perceived minimum cost.

Let the *demand function* be

$$t = t(z)$$

where z is the vector of expected minimum origin-to-destination costs. One would expect $t(z)$ to be monotonically decreasing, namely

$$(t'(z') - t''(z''))^T(z' - z'') < 0 \text{ for } z' \neq z''$$

Demand functions that are commonly encountered in the literature have an exponential form

$$t_w = t_0 \exp(-\beta z_w)$$

or a linear form

$$t_w = t_0 - \beta z_w$$

where t_0 is the maximum demand and β is the sensitivity to the expected cost. The exponential demand function fits conveniently with the logit choice model.

The presentation of the stochastic user equilibrium model with elastic demand given here follows Lam *et al.* (1995). Suppose first that link costs are fixed. The derivation of link choice proportions for the logit assignment model with fixed costs is discussed in Chapter 3. Define matrix $\mathbf{W}$ to be made up of the following elements

$$w_{mn} = \begin{cases} \exp(-\alpha c_{mn}) & \text{if a link connects node } m \text{ to node } n \\ 0 & \text{otherwise} \end{cases}$$

Bell (1995b) has shown that if the following series

$$N = W + W^2 + W^3 + \ldots = (I - W)^{-1} - I$$

is convergent then the proportion of traffic under the logit assignment model choosing link a, where link a connects node k to node l, is

$$p_{mna} = v_{mk} \exp(-\alpha c_{kl}) v_{ln} / v_{mn}$$

where v_{mk}, v_{ln} and v_{mn} are the corresponding elements of $\mathbf{N}$ and link a connects node $k \neq m$ to node $l \neq n$. All paths, including those with loops, are included.

In Section 6.3.1, it was noted that under the random utility interpretation of the logit model, the expected minimum cost of a trip from node m to node n is

$$z_{mn} = -(1/\alpha) \ln(v_{mn})$$

The estimated link flows are then
$$v = Pt(z)$$

where $\mathbf{P}$ is the matrix of link choice proportions.

The method of successive averages may now be applied as follows:

Method of successive averages algorithm for logit assignment with elastic demand

Step 1   (initialisation)
            $v \leftarrow 0$
            $c \leftarrow c(v)$
            $n \leftarrow 1$

Step 2   (method of successive averages)
            repeat
                calculate $N = (I - W)^{-1} - I$
                calculate P
                calculate z
                $v \leftarrow (1/n)Pt(z) + (1 - 1/n)v$
                $c \leftarrow c(v)$
                $n \leftarrow n + 1$
            until convergence.

As an example, consider the network in Fig. 5.7. The origin-to-destination demand functions are all assumed to have the same form, namely

$$t_{mn} = 10 \exp(-0.1 z_{mn})$$

Table 6.11 *Link cost functions, fitted flows and fitted costs*

Link	Origin node	Destination node	Slope	Intercept	Fitted flow	Fitted cost
1	1	5	0.2	2	5.877	3.175
2	2	7	0.2	2	5.877	3.175
3	5	6	0.5	2	5.877	4.939
4	7	8	0.5	2	5.877	4.939
5	5	7	1	2	1.031	3.031
6	7	5	1	2	1.031	3.031
7	6	8	0.3	2	1.740	2.522
8	8	6	0.3	2	1.740	2.522
9	6	3	0.2	2	5.877	3.175
10	8	4	0.2	2	5.877	3.175

Table 6.12 *Origin-to-destination fitted flows and expected minimum costs*

Origin	Destination	Fitted flow	Expected minimum cost
1	3	3.238	11.277
1	4	2.636	13.332
2	3	2.636	13.332
2	4	3.238	11.277

The maximum demand is 10 units (say, vehicles per second), which decreases as the expected cost increases, tending to zero. Linear cost–flow functions are assumed, with the intercept and slope terms given in Table 6.11. The fitted link flows and link costs are also given in Table 6.11. The fitted origin-to-destination flows are shown in Table 6.12 together with the expected minimum origin-to-destination costs.

6.8 SPACE–TIME NETWORKS

A more realistic description of traffic assignment, particularly in congested networks, requires the introduction of the time dimension. One way of doing this is to build the time dimension into the network. As described in Chapter 3, nodes in space–time networks constitute points not just in space but also in time. The links are assumed to have fixed capacities. When traffic cannot exit a link in the current interval it is queued to leave the link in a later interval. The queuing process is represented by links connecting a node to itself in a later interval.

In the case of deterministic user equilibrium, a difficulty with the *first-in, first-out* (FIFO) principle has been identified. This difficulty extends to stochastic user equilibrium assignment. FIFO leads to a non-convex set of feasible path flows (Carey, 1992). While queues obviously do exhibit a FIFO discipline, this is not so for traffic on links where overtaking is possible, so the desirability of a *rigid* enforcement of FIFO is questionable. However, without *any* FIFO discipline, an equivalent optimisation method would give unfettered priority to those streams that would result in the largest reductions to the objective function. The realism of the resulting assignment is questionable.

In the case of stochastic user equilibrium assignment, where some sub-optimal behaviour is permitted, all paths are used (although with decreasing probability as the cost of the path increases). The model would therefore suggest that some traffic opts to stay in queues longer than necessary, which would of course constitute unrealistic behaviour. Required is a model which allows the spreading of traffic across paths in the spatial domain but forces FIFO-like behaviour in the temporal domain (particularly at junctions). A model which goes some way towards this is described in the remainder of this section.

Noting that queues built up in one period must be processed in subsequent periods, a potentially useful compromise is the approximation of a dynamic assignment by a sequence of steady state equilibria linked across time by the queues. Equilibrium delay is a cost penalty required to bring demand into line with a fixed supply of infrastructure. However, underlying the notion of delay is some form of queuing, so a natural extension would be to translate equilibrium delays into equilibrium queues and carry the queues across periods. By explicitly introducing queues, the temporary overloading of links can be represented.

The equilibrium queue is equal to the equilibrium delay times the service rate. For example, if the equilibrium delay is 10 seconds and the service rate is 0.5 vehicles per second, the equilibrium queue will be 5 vehicles. Let $\mathbf{S}$ be a diagonal matrix of link service rates

$$\mathbf{S} = \begin{bmatrix} s_1 & & \\ & s_2 & \\ & & \ldots \end{bmatrix}$$

then

$$\mathbf{d}_t = \mathbf{S}^{-1}\mathbf{q}_t$$

where $\mathbf{d}_t$ is a vector of equilibrium link delays and $\mathbf{q}_t$ is a vector of equilibrium link queues.

The link capacity constraints would then have the form of the following complementary slackness conditions:

$$\mathbf{s} - \mathbf{A}\mathbf{h}_t - \mathbf{q}_{t-1} + \mathbf{q}_t \geqslant 0$$

$$\mathbf{q}_t \geqslant 0$$

$$(\mathbf{s} - \mathbf{A}\mathbf{h}_t - \mathbf{q}_{t-1} + \mathbf{q}_t)^{\mathrm{T}}\mathbf{q}_t = 0$$

where subscript t denotes a period, $\mathbf{q}_{t-1}$ is the (given) vector of queues at the end of period $t-1$ and $\mathbf{q}_t$ is the (variable) vector of queues at the end of period t. The demand for the links $\mathbf{A}\mathbf{h}_t$ can therefore exceed their steady state capacities $\mathbf{s}$ if the equilibrium queues are growing between periods, namely if $\mathbf{q}_t - \mathbf{q}_{t-1} > 0$. Note that the macroscopic effects of FIFO are captured in these constraints, because the queues carried over are processed before the new arrivals (by reducing the effective capacity of the corresponding links), and new queues are only formed when capacities are exceeded.

The following is a simplified version of the model proposed by Lam *et al.* (1995). Consider the following objective function:

$$f(\mathbf{h}_t, \mathbf{q}_t) = \mathbf{h}_t^{\mathrm{T}}(\ln(\mathbf{h}_t) - 1) + \alpha\mathbf{h}_t^{\mathrm{T}}\mathbf{A}^{\mathrm{T}}\mathbf{c} + 0.5\alpha\mathbf{q}_t^{\mathrm{T}}\mathbf{S}^{-1}\mathbf{q}_t$$

which is minimised subject to trip table constraints

$$\mathbf{t}_t = \mathbf{B}\mathbf{h}_t$$

and link capacity constraints

$$\mathbf{s} \geqslant \mathbf{A}\mathbf{h}_t + \mathbf{q}_{t-1} - \mathbf{q}_t$$

The Hessian of $f(\mathbf{h}_t, \mathbf{q}_t)$ is positive definite, so the objective function is strictly convex. The constraints are also convex, so the solution to the minimisation problem is unique in the primal variables $\mathbf{h}_t$ and $\mathbf{q}_t$. The Lagrangian equation has the form

$$L(\mathbf{h}_t, \mathbf{q}_t, \mathbf{u}_t, \mathbf{z}_t) = f(\mathbf{h}_t, \mathbf{q}_t) + \mathbf{z}_t^T(\mathbf{t}_t - \mathbf{B}\mathbf{h}_t) - \mathbf{u}_t^T(\mathbf{s} - \mathbf{A}\mathbf{h}_t - \mathbf{q}_{t-1} + \mathbf{q}_t)$$

The optimality conditions for $\mathbf{q}_t$ identify the dual variables $\mathbf{u}_t$ as being equal to the equilibrium delays times α, since

$$\mathbf{S}^{-1}\mathbf{q}_t^*\alpha = \mathbf{u}_t^* = \mathbf{d}_t^*\alpha$$

The optimality conditions for $\mathbf{h}_t$ are

$$\ln(\mathbf{h}_t^*) = -\mathbf{g}_t\alpha - \mathbf{A}^T\mathbf{u}_t^* + \mathbf{B}^T\mathbf{z}_t^*$$

which, together with the trip table constraints and the interpretation of $\mathbf{u}_t^*$ as the equilibrium delays times α, give the logit path choice model. The optimality conditions for $\mathbf{u}_t^*$ are

$$\mathbf{s} - \mathbf{A}\mathbf{h}_t^* - \mathbf{q}_{t-1}^* + \mathbf{q}_t^* \geqslant \mathbf{0}$$

$$\mathbf{u}_t^* \geqslant \mathbf{0}$$

$$\mathbf{u}_t^{*T}(\mathbf{s} - \mathbf{A}\mathbf{h}_t^* - \mathbf{q}_{t-1}^* + \mathbf{q}_t^*) = 0$$

Noting that a positive element of $\mathbf{u}_t$ corresponds to a positive element of $\mathbf{q}_t$ and a zero element of $\mathbf{u}_t$ corresponds to a zero element of $\mathbf{q}_t$, these optimality conditions give the required capacity–queue complementary slackness conditions given earlier.

An examination of the objective function reveals that, while trips are encouraged to spread spatially across paths by the entropy-related term $(\mathbf{h}_t^T(\ln(\mathbf{h}_t) - 1)$, queuing is *discouraged* by the quadratic term $0.5\alpha\mathbf{q}_t^T\mathbf{S}^{-1}\mathbf{q}_t$. Although temporary overloading of the network can be accommodated through queue formation, FIFO is not an issue. Since the role of the queue in one period is to reduce the capacity of the link in the subsequent period, the composition of the queue in terms of user class, origin or destination is not relevant.

This model may be fitted by either the method of successive averages (as proposed by Lam *et al.*, 1995) or by a modified version of the iterative balancing algorithm.

Method of successive averages for logit assignment

Step 1   (initialisation)
$$\mathbf{u}_t \leftarrow \mathbf{0}$$
$$\mathbf{z}_t \leftarrow \mathbf{0}$$
$$\mathbf{q}_t \leftarrow \mathbf{0}$$
$$n \leftarrow 1$$

Step 2   (successive averaging)
repeat
      for all trip tables elements w
$$\ln(\mathbf{h}_t)/\alpha \leftarrow -\mathbf{A}^T\mathbf{c} - \mathbf{A}^T\mathbf{u}_t + \mathbf{B}^T\mathbf{z}_t$$
$$\text{if } t_{w,t} \neq \sum_{\text{all } j \in P(w)} b_{wj}h_{j,t} \text{ then}$$

$$z_{w,t} \leftarrow z_{w,t} + \ln(t_{w,t}) - \ln\left(\sum_{\text{all } j \in P(w)} b_{wj}h_{j,t}\right)$$

```
for all links i
    ln(h_t) ← α − A^T c − A^T u_t + B^T z_t
    v_{i,t} ← Σ a_{ij} h_{j,t}
            all j
    if s_i < v_{i,t} + q_{i,t−1} then q_{i,t}* ← v_{i,t} + q_{i,t−1} − s_i
                  else q_{i,t}* ← 0
    q_{i,t} ← (1/n)q_{i,t}* + (1 − 1/n)q_{i,t}
    u_{i,t} ← (1/n)q_{i,t}/s_i
  n ← n + 1
until convergence.
```

In Step 2 of this simple but effective algorithm, the z variables are first calculated so that the pre-specified trip table t is reproduced. Then for each link i, a comparison of the current demand with the capacity leads to the definition of an auxiliary queue. This is then averaged with the current queue and the corresponding link delay is determined. The use of the method of successive averages can also allow for increasing link cost functions and multiple user classes (see Lam *et al.*, 1995).

The application of iterative balancing is not so simple.

Iterative balancing algorithm for logit assignment

Step 1 (initialisation)

```
    z_t ← 0
    q_t ← Max{0, q_{t−1} − s_t + 0.0001}
    u_t ← s^{−1} q_t α
```

Step 2 (iterative balancing)

```
    repeat
        for all origin—destination pairs k
            ln(h_t) ← −αA^T c − A^T u_t + B^T z_t
            if t_{w,t} ≠ b_w^T h_t then z_{w,t} ← z_{w,t} + ln(t_{w,t}) − ln(b_w^T h_t)
        for all links i
            ln(h_t) ← −αA^T c − A^T u_t + B^T z_t
            if s_i + q_{i,t} − q_{i,t−1} < a_i^T h_t then
                m_i ← (1 + s_i/α(s_i − q_{i,t−1} + q_{i,t}))
                u_{i,t} ← u_{i,t} − (ln(s_i − q_{i,t−1} + q_{i,t}) − ln(a_i^T h_t))/m_i
            q_{i,t} ← s_i u_{i,t}/α
    until convergence.
```

As in the method of successive averages, Step 2 starts by calculating the z variables so that the trip table t is reproduced. If the current as well as the past queue on link i were *fixed*, then the adjustment

$$u_{i,t} \leftarrow u_{i,t} - \ln(s_i - q_{i,t-1} + q_{i,t}) + \ln(a_i^T h_t)$$

(made only when the current demand plus the past queue exceeds the capacity plus the current queue) would lead to a convergent algorithm, because as shown in Chapter 3 the direction of change is an ascent direction and the size of the change is not sufficient to take the dual variable $u_{i,t}$ beyond its optimal value (the *ascent* does not become a *descent*). However, in this case

$$q_{i,t} = s_i u_{i,t}/\alpha$$

so the ascent could well become a descent if the above adjustment were made.

Let $u_{i,t}{}^a$ be the dual variable *after* the adjustment and $u_{i,t}$ the dual variable *before* the adjustment, and set

$$q_{i,t}^a = s_i u_{i,t}^a / \alpha$$

Then the non-negative adjustment

$$u_{i,t}{}^a \leftarrow u_{i,t} + \text{Max}\{-\ln(s_i - q_{i,t-1} + q_{i,t}{}^a) + \ln(\mathbf{a}_i^T \mathbf{h}_t), 0\}$$

would lead to a convergent algorithm since all such adjustments to the dual variables involve ascents only. Note that since the log function is concave

$$\ln(s_i - q_{i,t-1} + q_{i,t}^a) < \ln(s_i - q_{i,t-1} + q_{i,t}) + (q_{i,t}^a - q_{i,t})/(s_i - q_{i,t-1} + q_{i,t})$$

Hence by substitution in the above non-negative adjustment

$$u_{i,t}^a - u_{i,t} > -\ln(s_i - q_{i,t-1} + q_{i,t}) - (q_{i,t}^a - q_{i,t})/(s_i - q_{i,t-1} + q_{i,t}) + \ln(\mathbf{a}_i^T \mathbf{h}_t)$$

or

$$(u_{i,t}^a - u_{i,t}) > (-\ln(s_i - q_{i,t-1} + q_{i,t}) + \ln(\mathbf{a}_i^T \mathbf{h}_t))/m_i$$

where the modification factor is

$$m_i = (1 + s_i/\alpha(s_i - q_{i,t-1} + q_{i,t})) > 1$$

Thus Step 2 of the procedure makes smaller positive adjustments than the maximum possible while still ascending. Therefore *no* descents are made. While $q_{i,t}^a$ is not equal to $q_{i,t}$, ascents continue to be made, so the algorithm converges. In the case of the example presented here, convergence to three decimal places for both the primal and dual variables occurred within ten iterations.

When a sequence of time periods is considered, the above algorithm may be applied recursively. Starting with $\mathbf{q}_0$, $\mathbf{q}_1$ may be estimated, then $\mathbf{q}_2$ may be estimated from $\mathbf{q}_1$, etc.

Returning to the example network shown in Fig. 5.5 and the link costs and capacities given in Table 5.3, the model was applied recursively over six periods. For the first three periods there was a demand of 4 units of flow (say vehicles per second) and for the second three periods a demand of 1 unit of flow. Table 6.13 shows the flows and queues on links 1 and 5 (none of the other links had queues). Regarding link 1, the build up of the queue during the first three intervals is given by the difference between the flow and the capacity (one unit of flow). From the fourth period, the queue declines at a rate given by the difference between the capacity and the flow. The build up of the queue on link 5 is also given by the difference between the flow and the capacity (two units of flow). However, the capacity is sufficient for the queue to disappear entirely in the fourth interval. Having disappeared, the demand is insufficient for it to reform.

The fitted path flows are determined by the path costs, the trip table and the constraints. It is therefore instructive to look at the sensitivities of the fitted path flows to these inputs. The Tobin and Friesz (1988) approach to sensitivities may be applied here. Note that

$$\Delta \mathbf{u}_{a,t} = \mathbf{S}^{-1} \Delta \mathbf{q}_{a,t} \alpha$$

Table 6.13 *Queue and flow evolution over six periods*

Period	Link 1 Flow	Link 1 Queue	Link 5 Flow	Link 5 Queue	O-D flow
1	2.16	1.16	2.27	0.27	4
2	2.05	2.21	2.26	0.53	4
3	1.96	3.17	2.26	0.79	4
4	0.50	2.67	0.57	0	1
5	0.51	2.18	0.57	0	1
6	0.53	1.70	0.57	0	1

where subscript a denotes the set of links with active capacity constraints. At the optimum, the following relationships hold for small changes in the primal variables, $\Delta\mathbf{h}$ and $\Delta\mathbf{q}_a$, and the dual variables $\Delta\mathbf{z}$

$$\Delta\mathbf{g}\alpha = \mathbf{D}_t\Delta\mathbf{h}_t + \mathbf{B}^T\Delta\mathbf{z}_t - \mathbf{A}_a^T\mathbf{S}_a^{-1}\Delta\mathbf{q}_{a,t}\alpha$$

$$\Delta\mathbf{t}_t = \mathbf{B}\Delta\mathbf{h}_t$$

$$\Delta\mathbf{s}_a = \mathbf{A}_a\Delta\mathbf{h}_t - \Delta\mathbf{q}_{a,t}$$

where $\mathbf{D}_t$ is a diagonal matrix with elements $-1/h_{i,t}^*$ on the principal diagonal. Small changes in the path costs, the trip table and the active capacity constraints are related to small changes in the primal and dual variables as follows:

$$\begin{bmatrix} \Delta\mathbf{g} \\ \Delta\mathbf{t}_t \\ \Delta\mathbf{s}_a \end{bmatrix} = \begin{bmatrix} \mathbf{D}_t/\alpha & \mathbf{B}^T/\alpha & -\mathbf{A}_a^T\mathbf{S}_a^{-1} \\ \mathbf{B} & 0 & 0 \\ \mathbf{A}_a & 0 & -\mathbf{I} \end{bmatrix} \begin{bmatrix} \Delta\mathbf{h}_t \\ \Delta\mathbf{z}_t \\ \Delta\mathbf{q}_{a,t} \end{bmatrix}$$

The matrix between the two perturbation vectors is the Jacobian of the system. The inverse of this Jacobian provides the required sensitivities, *provided the constraints included are linearly independent*. The inverse of this Jacobian, represented in block form below, gives the sensitivities.

$$\begin{bmatrix} \Delta\mathbf{h}_t \\ \Delta\mathbf{z}_t \\ \Delta\mathbf{q}_{a,t} \end{bmatrix} = \begin{bmatrix} \mathbf{J}_{11} & \mathbf{J}_{12} & \mathbf{J}_{13} \\ \mathbf{J}_{21} & \mathbf{J}_{22} & \mathbf{J}_{23} \\ \mathbf{J}_{31} & \mathbf{J}_{32} & \mathbf{J}_{33} \end{bmatrix} \begin{bmatrix} \Delta\mathbf{g} \\ \Delta\mathbf{t}_t \\ \Delta\mathbf{s}_a \end{bmatrix}$$

Hence assuming that the active capacity constraints remain active and the inactive capacity constraints remain inactive, the perturbations have the following effect:

$$\Delta\mathbf{h}_t^* = \mathbf{J}_{11}\Delta\mathbf{g} + \mathbf{J}_{12}\Delta\mathbf{t}_t + \mathbf{J}_{13}\Delta\mathbf{s}_a$$

Variances and covariances may be obtained from the sensitivities. In particular, when path costs and link capacities remain unaltered,

$$\mathrm{Var}\{\mathbf{h}_t^*\} \approx \mathbf{J}_{12}\mathrm{Var}\{\mathbf{t}_t\}\mathbf{J}_{12}^T$$

Consider the network portrayed in Fig. 7.1. In order to simulate the build up and decline of queues over the peak period, the trip tables given in Table 6.14 were input. The value of the dispersion parameter α was taken to be 0.1.

The link flows and capacities are shown in Table 6.15. The link flows here are calculated from the path flows and represent the *demand* for the link. The capacities of links

Table 6.14 *Trip tables for eight periods*

O-D	Orgin-destination demand by period [units of flow]							
	1	2	3	4	5	6	7	8
A to C	2	5	6	5	4	3	2	1.5
A to D	2	5	6	5	4	3	2	1.5
B to C	2	5	6	5	4	3	2	1.5
B to D	2	5	6	5	4	3	2	1.5

Table 6.15 *Link capacities and demands*

Link	Capacity	Link demand by period [units of flow]							
		1	2	3	4	5	6	7	8
1	8	4	10	12	10	8	6	4	3
2	8	4	10	12	10	8	6	4	3
3	3	1.813	4.421	5.110	4.169	3.316	2.509	1.712	1.323
4	3	1.813	4.421	5.110	4.169	3.316	2.509	1.712	1.323
5	10	1.515	3.793	4.565	3.811	3.051	2.286	1.521	1.138
6	10	1.515	3.793	4.565	3.811	3.051	2.286	1.521	1.138

1 to 4 are exceeded in the second period. Demand falls below capacity on links 1 and 2 (which cannot be avoided) and links 3 and 4 (which can be avoided) in period 6.

The evolution of the queues are shown in Table 6.16. The queue falls to zero on links 1 and 2 but persists on links 3 and 4, due to the 'suppressed demand' from alternative higher cost paths not using links 3 and 4.

The sensitivities to path cost for period eight are shown in Table 6.17. As a result of the trip table constraints, the effect of path cost changes is limited to the origin–destination pair to which the path belongs. Note that the row and column sums of the sensitivities are zero.

The sensitivities of path flows to the trip table and the capacities of links 3 and 4 in the eighth period are shown in Table 6.18. Note that the column sums for each trip table element is one, indicating that flow is conserved. The column sums for the link constraints are zero as a result of the link capacity constraints.

The time-dependent methods presented so far have been based on deterministic queuing relationships of the form

$$q_t = \text{Max}\{\mathbf{A}\mathbf{h}_t^* + \mathbf{q}_{t-1}^* - \mathbf{s}, 0\}$$

Table 6.16 *Link costs and queues*

Link	Cost	Link queue by period [units of flow x period]							
		1	2	3	4	5	6	7	8
1	5	0	2	6	8	8	6	2	0
2	5	0	2	6	8	8	6	2	0
3	2	0	1.421	3.531	4.700	5.016	4.526	3.238	1.561
4	2	0	1.421	3.531	4.700	5.016	4.526	3.238	1.561
5	5	0	0	0	0	0	0	0	0
6	5	0	0	0	0	0	0	0	0

Table 6.17 *Sensitivity of path flow to path cost in period eight*

Path	Links	Sensitivity to change of cost on path [cost units]							
		1	2	3	4	5	6	7	8
1	1	−0.032	0.032	0	0	0	0	0	0
2	2,3,6	0.032	−0.032	0	0	0	0	0	0
3	2	0	0	−0.032	0.032	0	0	0	0
4	1,4,5	0	0	0.032	−0.032	0	0	0	0
5	2,3	0	0	0	0	−0.036	0.036	0	0
6	1,5	0	0	0	0	0.036	−0.036	0	0
7	1,4	0	0	0	0	0	0	−0.036	0.036
8	2,6	0	0	0	0	0	0	0.036	−0.036

Table 6.18 *Sensitivity of path flow to origin-destination demand and link capacity*

Path	Links	Origin-destination pair				Link	
		A to C	A to D	B to C	B to D	3	4
1	1	0.683	0.006	0	0	−0.011	0
2	2,3,6	0.317	−0.006	0	0	0.011	0
3	2	0	0	0.006	0.683	0	−0.011
4	1,4,5	0	0	−0.006	0.317	0	0.011
5	2,3	−0.004	0.555	0	0	0.012	0
6	1,5	0.004	0.455	0	0	−0.012	0
7	1,4	0	0	0.555	−0.004	0	0.012
8	2,6	0	0	0.455	0.004	0	−0.012

where the maximum operator is applied to each pair of elements. This leaves out of account both the random component of delay as well as the uniform component of delay under traffic signal control. Kimber and Hollis (1979) have presented more realistic time-dependent delay expressions that relate in principle to any kind of junction (see Chapter 4). These can be conveniently incorporated into the method of successive averages as follows:

Method of successive averages for logit assignment

Step 1   (initialisation)
$$\mathbf{u}_t \leftarrow \mathbf{0}$$
$$\mathbf{z}_t \leftarrow \mathbf{0}$$
$$\mathbf{q}_t \leftarrow \mathbf{0}$$
$$n \leftarrow 1$$

Step 2   (successive averaging)
repeat
for all origin—destination pairs w
$$\ln(\mathbf{h}_t)/\alpha \leftarrow -\mathbf{A}^T\mathbf{c} - \mathbf{A}^T\mathbf{u}_t + \mathbf{B}^T\mathbf{z}_t$$
if $t_{w,t} \neq \sum_{\text{all } j \in P(w)} b_{wj}h_{j,t}$ then
$$z_{w,t} \leftarrow z_{w,t} + \ln(t_{w,t}) - \ln(\sum_{\text{all } j \in P(w)} b_{wj}h_{j,t})$$
for all links i
$$\ln(\mathbf{h}_t)/\alpha - \mathbf{A}^T\mathbf{c} - \mathbf{A}^T\mathbf{u}_t + \mathbf{B}^T\mathbf{z}_t$$
$$v_{i,t} \leftarrow \sum_{\text{all } j} a_{ij}h_{j,t}$$
$$q_{i,t}, u_{i,t}{}^* \leftarrow \text{Kimber and Hollis}(1979) \leftarrow v_{i,t}, s_{i,t}, q_{i,t-1}$$
$$u_{i,t} \leftarrow (1/n)u_{i,t}{}^* + (1 - 1/n)u_{i,t}$$
$$n \leftarrow n + 1$$
until convergence.

While deterministic queuing relationships are probably the most convenient to use when little is known about the links, the Kimber and Hollis (1979) queue and delay expressions should yield more accurate assignment results.

6.9 DISCUSSION

Stochastic user equilibrium assignment has a number of attractions. The behavioural basis is more realistic than that of deterministic user equilibrium assignment. For the study of changes to trip-making behaviour in response to traveller information services, the allowance for less than perfect information is clearly helpful. Moreover, the introduction of a smooth demand curve (response surface) leads to uniqueness in terms of the path flows, a great practical advantage.

There are a number of directions in which future work on stochastic user equilibrium assignment can develop. One interesting avenue is the introduction of *multiple user classes* to reflect different levels of trip-maker information. In the case of the logit model, the level of information is reflected in the value of the dispersion parameter. Low values of the dispersion parameter reflect relative insensitivity to path cost and therefore a lower level of trip-maker information. As the dispersion parameter is increased, a deterministic user equilibrium assignment is approached.

Another interesting enhancement is the introduction of a time dimension. So far, dynamic traffic assignment has yet to be achieved satisfactorily by network flow programming, because of the lack of convexity caused by the FIFO requirement. A number of simulation models exist, in particular the CONTRAM model (Leonard *et al.*, 1978). A compromise approach is suggested here, which allows the application of convex programming methods. The queues at the end of one period affect the capacities of the effected links in the subsequent period. The macroscopic effects of FIFO are enforced, since the queues existing at the start of the period are processed *before* the new arrivals, and new queues are formed *only* when the capacities are reached. The model is, however, essentially steady state, as all the trips in the trip table are assumed to be fully realised within the period, even though the queues on links at the end of each period represent incomplete trips (In inconsistency in the assumptions). The queues in the model are demand-determined artefacts which maintain equilibrium between demand and supply.

Trip Table Estimation

7.1 INTRODUCTION

The traffic assignment methods discussed in Chapters 5 and 6 take as their input trip table information and output link statistics (like link flows, costs or delays) and possibly also path statistics (like path flows, costs or delays). In many transportation networks, however, it is difficult to obtain trip table information directly. Survey methods, like household, roadside or licence plate surveys, are often used for the purpose. For tolled facilities, like some express ways or public transport networks, electronic ticketing information provides an alternative source of trip table information. By contrast, link statistics, in particular link flows, are relatively easy to measure. Either manual methods or, in the case of vehicles, electronic detectors of various kinds, may be used. The problem of inferring trip table information from link measurements has therefore been a long-running theme of transportation network analysis.

A trip table estimator would appear to be the inverse of a traffic assignment method as the link statistics are input and trip table information is output. In general, however, additional data is required. Link measurements by themselves offer insufficient information for the reliable inference of trip tables. The additional data usually takes the form of prior information about the magnitude of elements of the trip table. The prior information emanates perhaps from an earlier survey or from a trip distribution model.

The first approach to the problem was inspired by introduction of the concept of *entropy* into quantitative methods for forecasting spatial interaction (see Wilson, 1967, on the use of entropy in urban and regional science). The hypothesis is that the trip table that maximised entropy subject to constraints imposed on it by link measurements would be the most likely trip table. An alternative to the entropy concept that leads essentially to the same models is the efficiency principle (Smith, 1987). However, the entropy approach makes a number of unattractive assumptions, in particular that link flows are measured without error and that traffic assignment is proportional (assignment is proportional when a doubling of the trip table would cause all link flows to double).

The next approach to the problem comes from the field of econometrics, and makes specific allowance for errors in the observations. The *generalised least squares* approach to trip table estimation, originally proposed by Cascetta (1984), incorporates not just link flow measurements and prior information about the sizes of origin-to-destination movements, but also their variance-covariance matrices. The assumption of proportional assignment is, however, retained. Lo *et al.* (1995) has looked at the effect of uncertainty in the link choice proportions.

The assumption of proportional assignment is justified in the rather exceptional case where all links are measured and where all link costs may therefore be estimated via link cost functions directly from the measured link flows. However, where networks are congested and where not all links are measured, it is necessary to introduce an assignment model.

An approach which allows non-proportional assignment, pursued by Yang et al. (1992), Yang (1995) and others, makes use of *bi-level programming* to combine deterministic user equilibrium assignment (the lower level problem) with, say, generalised least squares estimation (the upper level problem). This approach is computationally intensive but appears to work well in practice, when appropriately formulated.

The bi-level programming approach at least ensures consistency between the assignment assumptions made in estimation and those made in extrapolating the link flow measurements to the unmeasured links. However, a more direct approach would be to estimate the path flows directly from the link flow measurements incorporating the assignment assumptions in the estimator. The idea of *path flow estimation* was originally proposed by Sherali et al. (1994), but their approach required all links to be measured and assumed deterministic user equilibrium assignment. Bell et al. (1996) has proposed a path flow estimator based on stochastic user equilibrium assignment which does not require all links to be measured.

All the approaches so far mentioned are in effect *steady state*, requiring all trips started in the period modelled to be completed in the same period. Increasingly, interest is being shown in approaches which allow trips to span periods. Approaches based on time-dependent proportional assignment have been proposed (see, for example, Cascetta et al., 1993, and Ashok and Ben-Akiva, 1993). As mentioned in Chapters 5 and 6, dynamic assignment presents particular difficulties related to the FIFO principle and the non-convexity of the set of FIFO-feasible path flows. This chapter proposes an estimator that is time-dependent (rather than dynamic), does not assume proportional assignment and avoids difficulties relating to FIFO and convexity.

7.2 MAXIMUM ENTROPY

The concept of entropy has been introduced in Chapters 5 and 6. The number of different permutations of trips that give rise to a given trip table, referred to as the *entropy* of the trip table, is

$$E(\mathbf{t}) = \left(\sum_{\text{all } w} t_w\right)! \Bigg/ \prod_{\text{all } w} t_w!$$

where the sum and the product is over all admissible cells in the trip table (there may be some combinations of origins and destinations that are not considered, in particular movements within a zone). The trip table that maximises entropy would be the most likely if all permutations were equally probable, which of course is not the case as trips are not randomly allocated to cells. The prior information suggests that some cells of the trip table are likely to contain more trips than others, *ceteris paribus*.

Let **p** be the vector of probabilities that any trip selected at random will fall into any feasible cell of the trip table. Then the most likely trip table is that which maximises

$$L(\mathbf{t}) = \left(\prod_{\text{all } w} \exp(t_w \ln(p_w))\right)\left(\sum_{\text{all } w} t_w\right)! \Bigg/ \prod_{\text{all } w} t_w!$$

This is a *multinomial likelihood function*. The logarithm of $L(\mathbf{t})$ is

$$\ln(L(\mathbf{t})) = \sum_{\text{all } w} t_w \ln(p_w) + \ln\left(\left(\sum_{\text{all } w} t_w\right)!\right) - \sum_{\text{all } w} \ln(t_w!)$$

Sterling's approximation allows the removal of the factorials:

$$\ln(L(\mathbf{t})) = \sum_{\text{all } w} t_w \ln(p_w) + \left(\sum_{\text{all } w} t_w\right) \ln\left(\sum_{\text{all } w} t_w\right) - \sum_{\text{all } w} t_w \ln(t_w)$$

If $\mathbf{p}$ is based on a prior trip table $\mathbf{t}'$ and if the total number of trips in the prior trip table corresponds to the total number of trips in the estimated trip table, then

$$p_w = t_w' \bigg/ \sum_{\text{all } w} t_w'$$

and

$$\sum_{\text{all } w} t_w = \sum_{\text{all } w} t_w'$$

The logarithm of the likelihood function then reduces to

$$\ln(L(\mathbf{t})) = - \sum_{\text{all } w} t_w \ln(t_w'/t_w)$$

Hence the *equivalent optimisation problem* giving rise to the most likely trip table is

$$\text{Minimise } f(\mathbf{t}) = \sum_{\text{all } w} t_w \ln(t_w/t_w') \text{ subject to } \mathbf{v} = \mathbf{Pt}$$

where $\mathbf{P}$ is a matrix of *link choice proportions*. This is the estimator proposed by Willumsen (see Van Zuylen and Willumsen, 1980). It is assumed that link flows are measured accurately and that assignment is proportional. The Hessian of $f(\mathbf{t})$ is positive definite (provided the elements of the trip table are all positive; if not, the objective function is undefined) and the constraints are convex for given link choice proportions, so the solution to the optimisation problem is unique. At the optimum

$$\ln(\mathbf{t}^*) = \ln(\mathbf{t}') - \mathbf{1} + \mathbf{P}^T\mathbf{z}$$

where $\mathbf{z}$ are dual variables and $\mathbf{1}$ is a vector all of whose elements are 1.

Since the elements of $\mathbf{P}$ lie in the range 0 to 1 inclusive, the following iterative balancing scheme will converge to the solution (the proof of this is similar to that given in Chapter 3; the additional multiplicative constant term $\exp(\ln(\mathbf{t}') - \mathbf{1})$ does not affect the convergence of the procedure).

Iterative balancing algorithm for maximum entropy trip table estimation

Step 1 (initialisation)
     $\mathbf{z} \leftarrow \mathbf{0}$

Step 2 (iterative balancing)
     repeat

```
        for all constraints i
            t ← exp(ln(t') − 1 + PᵀZ)
            if vᵢ ≠ pᵢᵀt then zᵢ ← zᵢ + ln(vᵢ) − ln(pᵢᵀt)
    until convergence.
```

Note that $\mathbf{p}_i^T$ is the ith row of matrix $\mathbf{P}$. The speed of convergence depends on the size of the problem. If the constraints imposed by the traffic counts are mutually inconsistent, for example where the measured flow into an internal node is not equal to the measured flow out, the relevant dual variables will tend to either plus or minus infinity. Useful diagnostic facilities may therefore be included with the iterative balancing procedure. The maximum entropy trip table estimation method is widely used in practice, because of its simplicity of implementation and ease of interpretation.

Since

$$\nabla^2 f(\mathbf{t}^*) = \begin{bmatrix} 1/t_1^* & & \\ & 1/t_2^* & \\ & & \ddots \end{bmatrix}$$

the sensitivity expression developed in Chapter 3 may be applied here to obtain

$$d\mathbf{t}^*/d\mathbf{v} = (\nabla^2 f(\mathbf{t}^*))^{-1} \mathbf{P}^T (\mathbf{P}(\nabla^2 f(\mathbf{t}^*))^{-1} \mathbf{P}^T)^{-1}$$

This in turn yields variance–covariance expressions

$$\text{Var}\{\mathbf{t}^*\} = (d\mathbf{t}^*/d\mathbf{v})^T \text{Var}\{\mathbf{v}\}(d\mathbf{t}^*/d\mathbf{v})$$

This expression for the variances and covariances for the fitted values leaves out of account the effect of variation in the prior estimate of the trip table, $\mathbf{t}'$. To include this effect, the Tobin and Friesz (1988) approach may be applied as follows. At optimality, the following relationships hold for small changes in the primal and dual variables, $\Delta\mathbf{t}$ and $\Delta\mathbf{z}$ respectively:

$$\Delta \ln(\mathbf{t}') = \Delta \ln(\mathbf{t}) - \mathbf{P}^T \Delta\mathbf{z}$$

$$\Delta\mathbf{v} = \mathbf{P}\Delta\mathbf{t}$$

By a first order (linear) approximation

$$\Delta \ln(\mathbf{t}) = (\nabla^2 f(\mathbf{t}^*))\Delta\mathbf{t}$$

Hence small changes in the logarithm of the prior estimate of the trip table and the traffic counts are related to small changes in the primal and dual variables as follows.

$$\begin{bmatrix} \Delta \ln(\mathbf{t}') \\ \Delta\mathbf{v} \end{bmatrix} = \begin{bmatrix} \Delta^2 f(\mathbf{t}^*) & -\mathbf{P}^T \\ \mathbf{P} & 0 \end{bmatrix} \begin{bmatrix} \Delta\mathbf{t} \\ \Delta\mathbf{z} \end{bmatrix}$$

The matrix between the two perturbation vectors is the Jacobian of the system. The inverse of this Jacobian provides the required sensitivities. Writing the inverse Jacobian in the corresponding block form

$$\begin{bmatrix} \Delta\mathbf{t} \\ \Delta\mathbf{z} \end{bmatrix} = \begin{bmatrix} \mathbf{J}_{11} & \mathbf{J}_{12} \\ \mathbf{J}_{21} & \mathbf{J}_{22} \end{bmatrix} \begin{bmatrix} \Delta \ln(\mathbf{t}') \\ \Delta\mathbf{v} \end{bmatrix}$$

it can be shown that

$$\mathbf{J}_{11} = (\nabla^2 f(\mathbf{t}^*))^{-1}(\mathbf{I} - \mathbf{P}^T(\mathbf{P}(\nabla^2 f(\mathbf{t}^*))^{-1}\mathbf{P}^T)^{-1}\mathbf{P}(\nabla^2 f(\mathbf{t}^*))^{-1})$$

$$\mathbf{J}_{12} = (\nabla^2 f(\mathbf{t}^*))^{-1}\mathbf{P}^T(\mathbf{P}(\nabla^2 f(\mathbf{t}^*))^{-1}\mathbf{P}^T)^{-1}$$

$$\mathbf{J}_{21} = -(\mathbf{P}(\nabla^2 f(\mathbf{t}^*))^{-1}\mathbf{P}^T)^{-1}\mathbf{P}(\nabla^2 f(\mathbf{t}^*))^{-1}$$

$$\mathbf{J}_{22} = (\mathbf{P}(\nabla^2 f(\mathbf{t}^*))^{-1}\mathbf{P}^T)^{-1}$$

Hence the total effect on the fitted trip table is

$$\Delta\mathbf{t} = \mathbf{J}_{11}\Delta\ln(\mathbf{t}') + \mathbf{J}_{12}\Delta\mathbf{v}$$

so, assuming statistical independence between the prior estimates and the flow measurements, the variance–covariance matrix for the fitted trip table is given approximately as follows:

$$\text{Var}\{\mathbf{t}^*\} = \mathbf{J}_{11}\text{Var}\{\ln(\mathbf{t}')\}\mathbf{J}_{11}{}^T + \mathbf{J}_{12}\text{Var}\{\mathbf{v}\}\mathbf{J}_{12}{}^T$$

This requires a variance–covariance matrix for the log of the prior trip table. Given the non-negative nature of trip tables, an assumption that the elements of $\mathbf{t}'$ and $\mathbf{t}^*$ are lognormally distributed for the purposes of calculating confidence intervals makes sense. By a linear approximation

$$\text{Var}\{\ln(\mathbf{t}^*)\} = (\nabla^2 f(\mathbf{t}^*))\text{Var}\{\mathbf{t}^*\}(\nabla^2 f(\mathbf{t}^*))$$

Reference back to the expressions for $\mathbf{J}_{11}$ and $\mathbf{J}_{12}$ indicates that pre- and post-multiplication by $\nabla^2 f(\mathbf{t}^*)$ produces a significant simplification, since

$$\nabla^2 f(\mathbf{t}^*)\mathbf{J}_{11} = \mathbf{I} - \mathbf{P}^T(\mathbf{P}(\nabla^2 f(\mathbf{t}^*))^{-1}\mathbf{P}^T)^{-1}\mathbf{P}(\nabla^2 f(\mathbf{t}^*))^{-1}$$

$$\nabla^2 f(\mathbf{t}^*)\mathbf{J}_{12} = \mathbf{P}^T(\mathbf{P}(\nabla^2 f(\mathbf{t}^*))^{-1}\mathbf{P}^T)^{-1}$$

To illustrate the operation of the maximum entropy model, consider the network in Fig. 7.1. This has four trip table elements and six links. It is assumed that flow measurements are made on links 1 and 2, and that 10 units of flow (say vehicles per minute) are measured on each. The assumed link choice proportions are given in Table 7.1.

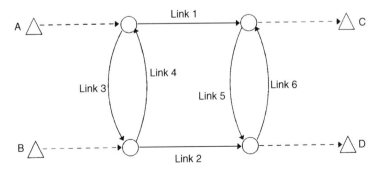

Figure 7.1 *An example network with six links*

Table 7.1 *Link choice proportions*

Link \ O-D pair	A to C	A to D	B to C	B to D
1	0.7	0.5	0.5	0.3
2	0.3	0.5	0.5	0.7

Table 7.2 *Post and prior estimates of the trip table*

O-D pair	Case 1 Prior	Case 1 Post	Case 2 Prior	Case 2 Post	Case 3 Prior	Case 3 Post
A to C	5	5	10	6.35	3.3	3.3
A to D	5	5	3.3	3.65	10	10
B to C	5	5	3.3	3.65	3.3	3.3
B to D	5	5	3.3	6.35	3.3	3.3

In this example, it took iterative balancing around 100 iterations to achieve an accuracy of four decimal places, so convergence is quite slow. Table 7.2 illustrates the effect of changes in prior information on the fitted trip table.

The sensitivities with respect to link flow measurements for Case 1 are given in Table 7.3. The estimate of flow from A to C is positively sensitive to the flow measurement on link 1 and negatively sensitive to the flow on link 2; the reverse is the case of the estimate of the flow from B to D. Conservation of flow requires that the row sums of this table be equal to one. Note that a unit increase in flow on link 1 causes more than a unit increase in the flow from A to C, because of the effect on the flow from B to D.

The sensitivities with respect to the logs of the elements of the prior trip table are given in Table 7.4. The fitted flows from A to D and from B to C are clearly more sensitive to the logs of the prior estimates than the fitted flows from A to C or from B to D, a fact born out by the results in Table 7.2. Due to the fact that *all* trips pass either link 1 or link 2, the row and column sums of this table are equal to zero.

It is worth noting the paucity of information used in the maximum entropy method. The only link information taken into account are link flow measurements; no information about link capacities or costs are made use of. Moreover, the assumption of proportional assignment is unrealistic in congested networks.

Table 7.3 *Sensitivity of the fitted trip table to link flows*

Link \ O-D pair	A to C	A to D	B to C	B to D
1	1.5	0.25	0.25	−1.0
2	−1.0	0.25	0.25	1.5

Table 7.4 *Sensitivity to the log of the prior trip table*

O-D pair \ O-D pair	A to C	A to D	B to C	B to D
A to C	1.25	−1.25	−1.25	1.25
A to D	−1.25	3.75	−1.25	−1.25
B to C	−1.25	−1.25	3.75	−1.25
B to D	1.25	−1.25	−1.25	1.25

7.3   A GENERALISED LEAST SQUARES

One disadvantage of the entropy maximising approach lies in the treatment of link flow observations as constraints and therefore as error-free. The generalised least squares approach emanating from econometrics provides a framework for allowing for errors from various sources. The method also yields standard errors for the trip table, thereby indicating the relative robustness of the fitted values. The method was first proposed by Cascetta (1984).

The equivalent optimisation problem has the form

$$\text{Minimise } f(\mathbf{t}) = 0.5(\mathbf{t} - \mathbf{t'})^T \mathbf{X}^{-1}(\mathbf{t} - \mathbf{t'}) + 0.5(\mathbf{v'} - \mathbf{Pt})^T \mathbf{Y}^{-1}(\mathbf{v'} - \mathbf{Pt})$$

The inputs are prior estimates of trip table flows ($\mathbf{t'}$), link flow measurements ($\mathbf{v'}$), variance–covariance matrices for the prior estimates and link flow measurements ($\mathbf{X}$ and $\mathbf{Y}$ respectively), and the matrix of link choice proportions ($\mathbf{P}$). As variance–covariance matrices are positive-definite, the objective function is convex and the minimum is uniquely given by

$$\nabla f(\mathbf{t^*}) = \mathbf{X}^{-1}(\mathbf{t^*} - \mathbf{t'}) - \mathbf{P}^T \mathbf{Y}^{-1}(\mathbf{v'} - \mathbf{Pt^*}) = \mathbf{0}$$

This yields the following linear estimator:

$$\mathbf{t^*} = (\mathbf{X}^{-1} + \mathbf{P}^T \mathbf{Y}^{-1} \mathbf{P})^{-1}(\mathbf{X}^{-1}\mathbf{t'} + \mathbf{P}^T \mathbf{Y}^{-1}\mathbf{v'})$$

Regarding sensitivities

$$\Delta \mathbf{t^*} = (\mathbf{X}^{-1} + \mathbf{P}^T \mathbf{Y}^{-1} \mathbf{P})^{-1}(\mathbf{X}^{-1}\Delta \mathbf{t'} + \mathbf{P}^T \mathbf{Y}^{-1}\Delta \mathbf{v'})$$

Let

$$\mathbf{E} = (\mathbf{X}^{-1} + \mathbf{P}^T \mathbf{Y}^{-1} \mathbf{P})^{-1}$$

then provided $\mathbf{t'}$ and $\mathbf{v'}$ are uncorrelated

$$\text{Var}\{\mathbf{t^*}\} = \mathbf{E}\mathbf{X}^{-1}\text{Var}\{\mathbf{t'}\}\mathbf{X}^{-1}\mathbf{E} + \mathbf{E}\mathbf{P}^T \mathbf{Y}^{-1}\text{Var}\{\mathbf{v'}\}\mathbf{Y}^{-1}\mathbf{PE} = \mathbf{E}\mathbf{X}^{-1}\mathbf{E} + \mathbf{E}\mathbf{P}^T \mathbf{Y}^{-1}\mathbf{PE}$$

The square root of the diagonal elements of Var $\{\mathbf{t^*}\}$ are the standard errors of the fitted values. As there are no constraints, there are no dual variables.

If the prior estimates and the link flow observations are both normally distributed with variances and covariances given by $\mathbf{X}$ and $\mathbf{Y}$ respectively, then there are two interpretations of the generalised least squares estimator. Firstly, it is equivalent to Bayesian estimation (see Maher, 1983). Secondly, it is equivalent to maximum likelihood estimation. Bell (1985) has shown that as $\mathbf{Y}$ tends to zero (meaning that the link flow measurements become error free), the generalised least squares model approximates the maximum entropy model, provided $\mathbf{X}$ is a diagonal matrix with the elements on the principal diagonal equal to the elements of $\mathbf{t^*}$. This is equivalent to the assumption that the link flow measurements are made without error but that the prior estimates are Poisson variates, because the variance of a Poisson variate is equal to its mean.

Unlike the maximum entropy model, there is nothing to prevent negative fitted values for the trip table being produced by the generalised least squares estimator. While negative values would reflect small real values, they are nonetheless counterintuitive. Bell (1991)

Table 7.5 *GLS estimates*

	Link flow measurement variance						ME estimates
	100	10	1	.1	.01	.001	
A to C	9.95	9.62	8.08	6.41	6.05	6.01	6.35
A to D	3.30	3.34	3.60	3.92	3.99	3.99	3.65
B to C	3.30	3.34	3.60	3.92	3.99	3.99	3.65
B to D	3.32	3.50	4.54	5.72	5.97	6.00	6.35
Link 1	11.27	11.12	10.62	10.12	10.01	10.00	10
Link 2	8.61	8.67	9.20	9.84	9.98	10.00	10

considers the introduction of non-negativity constraints for the fitted trip table and the effect these have on the variances and covariances. The method is straightforward to implement.

As an example, consider the network in Fig. 7.1, the link choice proportions in Table 7.1 and the prior estimates given in Case 2 of Table 7.2. For the prior trip table, assume that variances are equal to the prior estimates and that the covariances are zero. Table 7.5 shows the generalised least squares estimates for different flow measurement variances (each flow measurement has the same variance and the flow measurement covariances are zero). For comparison, the maximum entropy estimates are reproduced as well. As the flow measurement variance is progressively reduced, the fitted link flows move toward their measured values. Even with a very low variance, there is some discrepancy between the generalised least squares and maximum entropy fitted trip tables, illustrating that the one only *approximates* the other. Note that X was based on t'; the approximation would have been better if X had been based on t^*, but this would have required an iterative generalised least squares procedure.

7.4 BI-LEVEL PROGRAMMING

Both the maximum entropy and generalised least squares estimators described earlier make the assumption that traffic assignment is proportional, where P is the matrix of proportions. Where link cost functions are either increasing or right angle (the two cases considered in Chapters 5 and 6), the expected result of factoring up the trip table would *not* be proportional changes in link flows, unless link costs are effectively independent of link flows as might be the case in certain types of network when uncongested.

If all link flows are measured, then in principle all link costs may be derived from the link cost functions. The link choice proportions may then be derived from the assignment assumption. If deterministic user equilibrium assignment is assumed, then only least cost paths would be used (perhaps allowing some tolerance for measurement error). In this case, the path flows and therefore the link choice proportions are *not* uniquely defined (see Chapter 5). If stochastic user equilibrium assignment is assumed, then the demand for any path will depend on its cost according to some smooth, downward-sloping demand curve (or response surface). The path flows and therefore the link choice proportions are uniquely defined (see Chapter 6).

In general, link flow measurements are not available for all links, so there is no basis for calculating a fixed matrix of link choice proportions, unless link costs are independent of link flows. It is therefore usually necessary to integrate an assignment model more closely into the method of trip table estimation. One way of doing this is referred to as bi-level programming.

The traffic assignment model may be represented as a function with the trip table as input and link flows as output, namely

$$\mathbf{v} = A(\mathbf{t})$$

While given $\mathbf{t}$ it is possible to find $\mathbf{v}$, it is not possible to find $\mathbf{t}$ given $\mathbf{v}$, since in general the inverse of this function does not exist. It is therefore necessary to adopt an objective function, perhaps one of those previously considered, in order to arrive at an estimate of the trip table. Consider the following equivalent optimisation problem:

$$\text{Minimise } G(\mathbf{t}) = 0.5(\mathbf{t} - \mathbf{t}')^{\mathsf{T}}\mathbf{X}^{-1}(\mathbf{t} - \mathbf{t}') + 0.5(\mathbf{v}' - A(\mathbf{t}))^{\mathsf{T}}\mathbf{Y}^{-1}(\mathbf{v}' - A(\mathbf{t}))$$

For increasing link cost functions, the assignment function $A(\mathbf{t})$ is differentiable for both deterministic and stochastic user equilibrium assignment. Therefore

$$\nabla G(\mathbf{t}) = \mathbf{X}^{-1}(\mathbf{t} - \mathbf{t}') - \nabla A(\mathbf{t})^{\mathsf{T}}\mathbf{Y}^{-1}(\mathbf{v}' - A(\mathbf{t})$$

If the Jacobian of the assignment function $\nabla A(\mathbf{t})$ is independent of $\mathbf{t}$, then

$$\nabla^2 G(\mathbf{t}) = \mathbf{X}^{-1} + \nabla A(\mathbf{t})^{\mathsf{T}}\mathbf{Y}^{-1}\nabla A(\mathbf{t})$$

is positive definite, since $\mathbf{X}$ and $\mathbf{Y}$ are variance–covariance matrices, and there is a unique solution to the equivalent optimisation problem.

For both deterministic and stochastic user equilibrium assignment

$$\Delta\mathbf{v} = \mathbf{A}(\nabla^2 f(\mathbf{h}^*))^{-1}\mathbf{B}^{\mathsf{T}}(\mathbf{B}(\nabla^2 f(\mathbf{h}^*))^{-1}\mathbf{B}^{\mathsf{T}})^{-1}\Delta\mathbf{t}$$

where $\mathbf{A}$ is the link–path incidence matrix, $\mathbf{B}$ is the trip table-path incidence matrix and $f(\mathbf{h})$ is the objective function of the equivalent optimisation problem which yields the assignment function (see Chapters 5 and 6). Note that approximately at least

$$\Delta\mathbf{v} = \nabla A(\mathbf{t})\Delta\mathbf{t}$$

so by substitution

$$\nabla A(\mathbf{t}) = \mathbf{A}(\nabla^2 f(\mathbf{h}^*))^{-1}\mathbf{B}^{\mathsf{T}}(\mathbf{B}(\nabla^2 f(\mathbf{h}^*))^{-1}\mathbf{B}^{\mathsf{T}})^{-1}$$

In the case of deterministic user equilibrium assignment, $\mathbf{h}^*$ represents the solution corresponding to a given minimal set of paths and

$$\nabla^2 f(\mathbf{h}^*) = \mathbf{A}^{\mathsf{T}}\nabla\mathbf{c}(\mathbf{v})\mathbf{A}$$

so the Jacobian of the assignment function is constant (independent of $\mathbf{t}$) if the Jacobian of the link cost functions is constant, in other words if the link cost functions are linear. In the case of stochastic user equilibrium assignment,

$$\nabla^2 f(\mathbf{h}^*) = \mathbf{D} + \alpha\mathbf{A}^{\mathsf{T}}\nabla\mathbf{c}(\mathbf{v})\mathbf{A}$$

where $\mathbf{D}$ is a diagonal matrix whose jth element is $1/h_j{}^*$. This is not independent of $\mathbf{t}$.

In general, though, $G(\mathbf{t})$ is believed not to be convex, indicating that the problem of trip table estimation with non-proportional assignment falls outside the realms of convex programing.

Since the assignment function is itself usually the solution to an equivalent programing problem, an approach to such estimation problems is to define two optimisation problems, an upper level one and a lower level one. This is referred to as a *bi-level programming problem*. The upper level is conventionally the trip table estimation problem and the lower level the assignment problem. While both optimisation problems may be convex, the underlying problem is generally non-convex, in the sense that there are many local solutions where linear combinations of these solutions may not be feasible.

The conventional approach to solving bi-level programming problems is to pass parameters between the two levels in an iterative way, solving the upper and lower level problems in sequence. There is clearly a question as to whether such iterative procedures converge. In some cases, experience has been encouraging (Yang *et al.*, 1992).

Considering the case of generalised least squares estimation in combination with deterministic user equilibrium assignment, there are a number of ways of proceeding. One configuration would be

Upper level: $\mathbf{t}_{k+1} = (\mathbf{X}^{-1} + \mathbf{P}_k^T \mathbf{Y}^{-1} \mathbf{P}_k)^{-1} (\mathbf{X}^{-1} \mathbf{t}' + \mathbf{P}_k^T \mathbf{Y}^{-1} \mathbf{v}')$
                $k \leftarrow k + 1$

Lower level: Evaluate $\mathbf{P}_k$ by deterministic user equilibrium
                assignment

whereby the estimated trip table $\mathbf{t}$ is passed down and the matrix of link choice proportions $\mathbf{P}$ is passed up.

Yang (1995) found that a better alternative is to use the matrix of sensitivities $\nabla A(\mathbf{t})$ in the upper level problem rather than $\mathbf{P}$. A linear approximation to the assignment function at iteration k is then

$$\mathbf{v}' = \mathbf{v}_k + \nabla A(\mathbf{t}_k)(\mathbf{t} - \mathbf{t}_k)$$

Given $\mathbf{v}_k$ and $\mathbf{t}_k$, the following linear constraints

$$\mathbf{v}' - \mathbf{v}_k + \nabla A(\mathbf{t}_k)\mathbf{t}_k = \nabla A(\mathbf{t}_k)\mathbf{t}$$

replace

$$\mathbf{v}' = \mathbf{P}\mathbf{t}$$

on iteration k. This yields the following configuration:

Upper level: $\mathbf{t}_{k+1} \leftarrow (\mathbf{X}^{-1} + \nabla A(\mathbf{t}_k)^T \mathbf{Y}^{-1} \nabla A(\mathbf{t}_k))^{-1} (\mathbf{X}^{-1} \mathbf{t}' + \nabla A(\mathbf{t}_k)^T \mathbf{Y}^{-1}$
                $(\mathbf{v}' - \mathbf{v}_k + \nabla A(\mathbf{t}_k)\mathbf{t}_k))$
                $\mathbf{v}_{k+1} \leftarrow \mathbf{v}_k + \nabla A(\mathbf{t}_k)(\mathbf{t}_{k+1} - \mathbf{t}_k)$
                $k \leftarrow k + 1$

Lower level: Evaluate $\nabla A(\mathbf{t}_k)$ by deterministic user equilibrium
                assignment

where, as before, the estimated trip table is passed down, but this time the matrix of sensitivities is passed up.

As an example, consider the network given in Fig. 7.1. Each link is assumed to have the same cost function, namely

$$\text{cost} = 1 + (\text{flow})^2$$

On both links 1 and 2, flow measurements of 10 units (say vehicles per minute) are made, each with a variance of 10 units². The covariances are assumed to be zero. The prior estimates for the trip table flows from A to C, A to D, B to C and B to D are assumed to be 10, 4, 4, and 4 units respectively. The corresponding variances are likewise assumed to be 10, 4, 4 and 4 units² respectively, with zero covariances. Starting with the prior trip table, the deterministic user equilibrium assignment algorithm finds sensitivities for links 1 and 2 with respect to small changes in the trip table. The flows on links 1 and 2 are estimated to be 16.67 and 17.32 units respectively, which differ from the measured values. The sensitivities are passed to the generalised least squares algorithm, which estimates a new trip table. The new trip table is passed back to the deterministic user equilibrium assignment algorithm, etc. The sequence of link-trip table sensitivities is given in Table 7.6 and the sequence of fitted trip tables in Table 7.7. Convergence is achieved after 2 iterations (the results for iteration 3 were the same as those for iteration 2, to four decimal places).

For the calculation of the sensitivities using the method described in Chapter 5, it was necessary to restrict the set of paths to a minimal set corresponding to an extreme point. The network has eight paths, but only six of them can be linearly independent because there are six linearly independent links. In the second iteration of the bi-level programming method, the link flows and costs were as given in Table 7.8. The costs suggest that, while two paths would be used between A and D and between B and C, only one would be used be between A and C as well as between B and D. Consequently, the path via links 3, 2 and 6 as well as the path via links 4, 1 and 5 were excluded

Table 7.6   *Sensitivities for links 1 and 2 for 2 iterations*

Iteration	Link	A to C	A to D	B to C	B to D
1	1	0.6318	0.5108	0.3845	0.3845
1	2	0.3682	0.4892	0.6155	0.6155
2	1	0.6316	0.5106	0.3843	0.3843
2	2	0.3684	0.4894	0.6157	0.6157

Table 7.7   *Fitted trip table for 2 iterations*

Iteration	A to C	A to D	B to C	B to D
1	9.4281	3.8094	3.8492	3.8492
2	9.4282	3.8094	3.8493	3.8493

Table 7.8   *Link flows and costs*

	Link					
	1	2	3	4	5	6
Flow	10.69	10.25	3.22	0.67	0.66	3.25
Cost	115.27	105.99	11.36	1.45	1.44	11.57

from the link–path and trip table–path incidence matrices, $\mathbf{A}$ and $\mathbf{B}$ respectively, before calculation of the sensitivities. The consequence of *not* so doing would be that the matrix $\mathbf{A}^T\nabla\mathbf{c}(\mathbf{v})\mathbf{A}$ would be singular and therefore not invertible.

Experience suggests that the bi-level programming method with a generalised least squares upper level problem and a deterministic user equilibrium lower level problem converges well in practice.

In place of the generalised least squares upper level problem, a maximum entropy problem would be possible within the bi-level framework, for example

Upper level: $\mathbf{t}_{k+1} \leftarrow$ Minimise $f(\mathbf{t}) = \mathbf{t}^T(\ln(\mathbf{t}) - \ln(\mathbf{t}'))$
              subject to $\mathbf{v}' - \mathbf{v}_k + \nabla A(\mathbf{t}_k)\mathbf{t}_k = \nabla A(\mathbf{t}_k)\mathbf{t}$
              $\mathbf{v}_{k+1} \leftarrow \mathbf{v}_k + \nabla A(\mathbf{t}_k)(\mathbf{t}_{k+1} - \mathbf{t}_k)$
              $k \leftarrow k + 1$

Lower level: Calculate $\nabla A(\mathbf{t}_k)$

In place of the deterministic user equilibrium assignment lower level problem, a stochastic user equilibrium assignment problem could be substituted. However, in the case of stochastic user equilibrium assignment, there is a better approach altogether.

7.5 LINEAR PATH FLOW ESTIMATION

In Chapter 5, it has been shown that when links have fixed costs, a deterministic user equilibrium assignment may be found by solving a linear programming problem. The first path flow estimator, due to Sherali *et al.* (1994), made the assumption that *all* link costs are known, although they do suggest what to do when this is not the case. When all link costs are known, the least cost path for every trip table pair may be found by a shortest path algorithm (see Chapter 2 for a description of two such algorithms). Let $\mathbf{A}^*$ be the link-path incidence matrix corresponding to the least cost paths, then the following linear programing problem:

$$\text{Minimise } \mathbf{h}^T\mathbf{r} \text{ subject to } \mathbf{v}' = \mathbf{A}^*\mathbf{h}, \mathbf{h} \geqslant \mathbf{0}$$

where $\mathbf{r}$ is any vector of weights of compatible dimension and $\mathbf{v}'$ is the vector of measured link flows, yields an *extreme point* (a path flow vector with the minimum number of non-zero elements compatible with $\mathbf{v}'$; see Chapter 5). When $\mathbf{r}$ is replaced by $\mathbf{g}$, the vector of path costs, the Sherali estimates in its most basic form is obtained. Since this estimator requires knowledge of all link costs, it is for practical purposes not useful.

When flow measurements are available for some links, the link–path incidence matrix may be partitioned so that one block $\mathbf{A}'$ relates to the measured links and the other block $\mathbf{A}''$ relates to the unmeasured links. For the unmeasured links, right angle cost functions are assumed, with constant costs $\mathbf{c}''$ and capacities $\mathbf{s}''$. Let $\mathbf{z}$ be the minimum origin-to-destination costs, found by a shortest path algorithm. A deterministic user equilibrium assignment may be found by solving the following linear programming problem.

$$\text{Minimise } \mathbf{h}^T(\mathbf{g} - \mathbf{B}^T\mathbf{z}) \text{ subject to } \mathbf{v}' = \mathbf{A}'\mathbf{h}, \mathbf{s}'' \geqslant \mathbf{A}''\mathbf{h}, \mathbf{h} \geqslant \mathbf{0}$$

where

$$\mathbf{g}^T\mathbf{h} = \mathbf{c}^T\mathbf{A}\mathbf{h} = \mathbf{c}'(\mathbf{v}')^T\mathbf{A}'\mathbf{h} + \mathbf{c}''^T\mathbf{A}''\mathbf{h} = \mathbf{c}'(\mathbf{v}')^T\mathbf{v}' + \mathbf{c}''^T\mathbf{v}'' = \text{constant} + \mathbf{c}''^T\mathbf{v}''$$

Note that

$$g - B^T z \geqslant 0$$

so that if a trip is on a least cost path there is no advantage in changing path. The effect of the objective function is to fill the least cost paths first. At the optimum (see Chapter 3 on optimality conditions for convex programing problems)

$$g - B^T z = A'^T w - A''^T u + y, \; u \geqslant 0, \; y \geqslant 0$$

where w are the dual variables for the link flow measurements, u the dual variables for the capacity constraints and y the dual variables for the non-negativity conditions. The corresponding complementary slackness conditions for the dual variables are

$$w^T (v' - A'h) = 0$$

$$u \geqslant 0$$

$$s'' - A''h \geqslant 0$$

$$u^T (s'' - A''h) = 0$$

$$y \geqslant 0$$

$$h \geqslant 0$$

$$y^T h = 0$$

The dual variables u and w are the *shadow prices* of the links, representing the excess cost that a particular constraint or measurement imposes on a marginal trip-maker. Hence, the u variables may be interpreted as the deterministic user equilibrium delays for capacity constrained links and the w variables correspond to the excess cost incurred by the marginal measured unit of flow. In summary, the above linear programming problem yields deterministic user equilibrium path flows as well as the equilibrium flows and delays for the unmeasured, capacity constrained links (with a proviso mentioned later).

Consider the example in Fig. 7.1 when the link flow measurements, costs and capacities are as in Table 7.9. There are four least cost paths that do not use links 5 and 6. Two non-zero path flows (say 10 units of flow from A via link 1 to C and 10 units of flow from B via link 2 to D) are sufficient to reproduce the two link flow measurements. This solution constitutes an extreme point. However, a prior or target trip table may suggest that flow is expected from B to C and from A to D as well.

If a target trip table t' is available, the following linear programming problem minimises the total deviation from the target trip table subject to the measured link flows v', the capacities for the unmeasured links s'', and a *preference* for the least cost paths.

Minimise $h^T (g - B^T z) + \alpha 1^T (x^+ + x^-)$

Table 7.9 *Link flows, costs and capacities*

Link	1	2	3	4	5	6
Measurement	10	10				
Capacity			5	5	10	10
Cost	5	5	2	2	5	5

subject to $\mathbf{Bh} + \mathbf{x}^+ - \mathbf{x}^- = \mathbf{t}', \mathbf{v}' = \mathbf{A}'\mathbf{h}, \mathbf{s}'' \geqslant \mathbf{A}''\mathbf{h}, \mathbf{x}^+ \geqslant \mathbf{0}, \mathbf{x}^- \geqslant \mathbf{0}, \mathbf{h} \geqslant \mathbf{0}$

where $\mathbf{x}^+$ and $\mathbf{x}^-$ are non-negative slack variables and α is a weight applied to the total slackness (deviation from the target trip table). If, as in the above example, there are many vectors of path flows that are compatible with a deterministic user equilibrium, then one may wish to identify that one which is closest to a target trip table (in other words, the preference for least cost paths is *absolute*). This requires a two-phase approach, first finding the deterministic user equilibrium link flows (phase one) and then finding the path flows that are compatible with those link flows and closest to the target trip table (phase two). In the above example, however, the deterministic user equilibrium link flows are not unique because the cost functions for the unmeasured links are right angular.

Consider the network in Fig. 7.2. The link flows, costs and capacities are given in Table 7.10 for Case 1. The least cost path uses links 1 and 4, but the capacity of link 4 is insufficient for the measured flow on link 1. The capacity constraint on link 4 is therefore active, the marginal measured unit of flow incurs an excess cost of 3 units because it is forced to use link 2 and the constraint on link 4 imposes an excess cost of 3 units. Numbering the paths as in Table 7.11, then

$$(\mathbf{g} - \mathbf{B}^\mathsf{T}\mathbf{z})^\mathsf{T} = (0, 3, 3, 6) = (\mathbf{A}'^\mathsf{T}\mathbf{w} - \mathbf{A}''^\mathsf{T}\mathbf{u} + \mathbf{y})^\mathsf{T}$$

$$= (3, 3, 0, 0) - (3, 0, 3, 0) + (0, 0, 6, 6)$$

Thus $u_4 = 3$ units of cost and $u_1 = u_2 = u_3 = 0$ units of cost, $w_1 = 3$ units of cost and $w_2 = w_3 = w_4 = 0$ units of cost, and $y_3 = y_4 = 6$ units of cost and $y_1 = y_2 = 0$ units of cost. Paths 3 and 4 are not used because both y_3 and y_4 are positive. The solution path

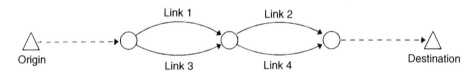

Figure 7.2 *Example network with 4 links*

Table 7.10 *Link flows, costs and capacities for Case 1*

Link	1	2	3	4
Measurement	10			
Capacity		10	10	5
Cost	2	5	5	2

Table 7.11 *Paths and excess costs*

Path	1	2	3	4
Links	1, 4	1, 2	3, 4	3, 2
Cost	4	7	7	10
Excess	0	3	3	6

flows are therefore $h_1{}^* = 5$ and $h_2{}^* = 5$. Note that, from the complementary slackness conditions, **u** has positive elements only where a capacity constraint is active, **w** has positive elements only where there is a link measurement, and **y** has positive elements only where paths are unused.

Case 2 is shown in Table 7.12. The measurement of 10 units of flow is now on link 2, which is not on a least cost path. These 10 units of flow would prefer to use link 1 rather than link 3, but there is insufficient capacity for all. The marginal measured unit therefore incurs an excess cost of 6 units.
Thus

$$(\mathbf{g} - \mathbf{B}^T\mathbf{z})^T = (0, 3, 3, 6) = (\mathbf{A}'^T\mathbf{w} - \mathbf{A}''^T\mathbf{u} + \mathbf{y})^T$$
$$= (0, 6, 0, 6) - (3, 3, 0, 0) + (3, 0, 3, 0)$$

This implies that paths 1 and 3 are *not* used and that paths 2 and 4 may be used. In this case, the solution path flows are $h_2{}^* = 5$ and $h_4{}^* = 5$. However, the flow measured on link 2 is not on link 4 presumably because of a capacity constraint on that link, suggesting that there is a total flow of 15 units from the origin to the destination. This extra flow is not generated by the path flow estimator because it would have no effect on the objective function. The path flow estimator therefore assigns the measured flows according to a deterministic user equilibrium, but makes no inferences about demands that are not measured. *It is therefore important that measurements are made on all least cost paths.*

Table 7.12 *Link flows, costs and capacities for Case 2*

Link	1	2	3	4
Measurement		10		
Capacity	5		10	5
Cost	2	5	5	2

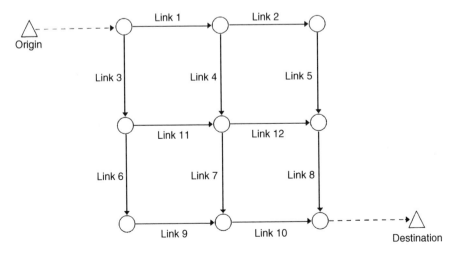

Figure 7.3 *An example network with 12 links*

Table 7.13 *Measured link flows, costs and capacities*

Link	1	2	3	4	5	6	7	8	9	10	11	12
Flow	5		5									
Capacity		5		2	5	5	2	10	5	10	2	2
Cost	2	5	5	2	5	5	2	5	5	2	2	2

Table 7.14 *Paths and their excess costs*

Path	Links	Excess cost	Flow
1	1, 4, 7, 10	0	2
2	1, 4, 12, 8	3	0
3	3, 11, 7, 10	3	0
4	3, 11, 12, 8	6	0
5	3, 6, 9, 10	9	5
6	1, 2, 5, 8	9	3

As a further example, consider the network in Fig. 7.3. The measured link flows, costs and capacities are given in Table 7.13. The paths and their excess costs are given in Table 7.14. Links 4 and 11 have active capacity constraints. In the case of link 4, the cost of the alternative route is 9 units more, so $u_4 = 9$. In the case of link 11, the cost of the alternative route is 6 units more, so $u_{11} = 6$. The marginal unit of flow on either link 1 or link 3 incurs an excess cost of 9 units. Thus

$$(g - B^T z)^T = (0, 3, 3, 6, 9, 9) = (A'^T w - A''^T u + y)^T$$

$$= (9, 9, 9, 9, 9, 9) - (9, 9, 6, 6, 0, 0) + (0, 3, 0, 3, 0, 0)$$

This indicates that paths 2 and 4 are *not* used. The configuration of the traffic flow measurements and capacities have ensured that, while paths 2 and 4 with excess costs of 3 and 6 respectively are not used, paths 5 and 6 with excess costs of 9 each are.

Using slack variables, errors in the traffic flow measurements may also be allowed for (see Sherali *et al.*, 1994). The approach is the same as with the target trip matrix. The slack variables allow the constraints imposed by link flow measurements to be relaxed, but the total slackness is penalised in the objective function.

The method as described so far requires the prior identification of paths, which in practical terms is a disadvantage. Sherali *et al.* (1994) discuss the problem of column generation. This topic is dealt with more fully in the next section.

7.6 LOG-LINEAR PATH FLOW ESTIMATION

In Chapter 6, it was shown that the stochastic user equilibrium assignment problem has a unique solution in terms of path flows. Furthermore, in the case of the logit route choice model, there is an equivalent optimisation problem due to Fisk (1980). The objective function has two terms, one which seeks to maximise the path entropy, and so *spread* the trips, and the other which seeks to *concentrate* the trips on least cost paths.

Consider the following problem

$$\text{Minimise } h^T(\ln(h) - 1) + \alpha g^T h \text{ subject to } v' = A'h, s'' \geqslant A''h$$

where as before the links have been partitioned into those with flow measurements $\mathbf{v}'$ and those with known capacities $\mathbf{s}''$. The corresponding link-path incidence matrices are $\mathbf{A}'$ and $\mathbf{A}''$. Non-negativity constraints on the vector of path flows $\mathbf{h}$ is implicit from the form of the objective function. Parameter α governs the amount of dispersion across sub-optimal paths (a larger α means less dispersion). The vector of path costs is $\mathbf{g}$ and this is made up of link costs. The total system cost (net of delay) has a constant and a variable component as follows:

$$\mathbf{g}^T\mathbf{h} = \mathbf{c}^T\mathbf{A}\mathbf{h} = \mathbf{c}'(\mathbf{v}')^T\mathbf{A}'\mathbf{h} + \mathbf{c}''^T\mathbf{A}''\mathbf{h} = \mathbf{c}'(\mathbf{v}')^T\mathbf{v}' + \mathbf{c}''^T\mathbf{v}'' = \text{constant} + \mathbf{c}''^T\mathbf{v}''$$

For measured links, costs are assumed to be known, while for unmeasured links, uncongested (or free-flow) costs are used. For unmeasured links, right angle cost functions are assumed. The right angle cost function has the practical advantage of only requiring two parameters; the free-flow cost and the capacity.

At the optimum (see Chapter 3 on optimality conditions for convex programing problems)

$$\ln(\mathbf{h}^*) + \alpha\mathbf{g} = \mathbf{A}'^T\mathbf{w}^* - \mathbf{A}''^T\mathbf{u}^*, \mathbf{u}^* \geqslant \mathbf{0}$$

where $\mathbf{w}$ are the dual variables for the link flow measurements and $\mathbf{u}$ the dual variables for the capacity constraints. The corresponding complementary slackness conditions are

$$\mathbf{w}^T(\mathbf{v}' - \mathbf{A}'\mathbf{h}) = 0$$

$$\mathbf{u} \geqslant \mathbf{0}$$

$$\mathbf{s}'' - \mathbf{A}''\mathbf{h} \geqslant \mathbf{0}$$

$$\mathbf{u}^T(\mathbf{s}'' - \mathbf{A}''\mathbf{h}) = 0$$

The dual variables $\mathbf{u}$ and $\mathbf{w}$ are the shadow prices for the unmeasured and measured links respectively, as is demonstrated in what follows.

Consider a path j with an active capacity constraint on link i. A small relaxation of this constraint causes a small quantity of flow to move to path j directly or indirectly from some other path j'. For path j, the objective function changes by

$$\Delta f(\Delta h_j) = (h_j^* + \delta)(\ln(h_j^* + \delta) - 1) - h_j^*(\ln(h_j^*) - 1) + \alpha\delta g_j$$

where δ represents a small positive amount of flow. By a linear approximation

$$(h_j^* + \delta)(\ln(h_j^* + \delta) - 1) - h_j^*(\ln(h_j^*) - 1) \approx \delta\ln(h_j^*)$$

so

$$\Delta f(\Delta h_j) \approx \delta\ln(h_j^*) + \alpha\delta g_j$$

For path j', the objective function changes by

$$\Delta f(\Delta h_{j'}) \approx -\delta\ln(h_{j'}^*) - \alpha\delta g_{j'}$$

The dual variable of the constraint indicates the rate at which the optimum value of the objective function reduces as the constraint is relaxed, so as δ tends to zero,

$$\ln(h_j^*/h_{j'}^*) + \alpha(g_j - g_{j'}) = u_i$$

If paths j and j' connect the same origin to the same destination, then

$$\ln(h_j^*/h_{j'}^*) = -\alpha(g_j - g_{j'}) - u_i$$

is a logit path choice model when u_i are interpreted as the stochastic user equilibrium link delay multiplied by α, since path j has one bottleneck at link i and path j' has no bottleneck *after* the transfer of δ trips. The model is therefore compatible with the logit assignment model presented in Chapter 6 and, as before, the elements of $\mathbf{u}$ are interpretable as the equilibrium link delays divided by α.

As indicated in Chapter 3, the elements of $\mathbf{u}$ are only unique if the corresponding constraints are linearly independent, which in turn implies that only linearly independent links should be included. The number of linearly independent links is equal to the total number of links minus the number of internal nodes, as noted earlier.

The interpretation of the elements of $\mathbf{w}$ is as follows. At the optimum

$$\ln(h_j^*) = -\alpha g_j - \sum_i a_{ij}'' u_i + \sum_i a_{ij}' w_i$$

where

$$\sum_i a_{ij}'' u_i = \text{total equilibrium delay for path } j \text{ times } \alpha$$

and

$$\sum_i a_{ij}' w_i = \text{total shadow price (or more generally the implicit value of the measured demand) for path } j \text{ times } \alpha$$

The dual variables may be estimated either by the method of successive averages or by iterative balancing, both described in Chapter 3, provided the measured and unmeasured link–path incidence matrices, $\mathbf{A}'$ and $\mathbf{A}''$, are available. The following algorithm, presented in Bell (1995c), solves for the dual variables by iterative balancing.

Iterative balancing algorithm for logit path flow estimator

Step 1   (initialisation)
            $\mathbf{u} \leftarrow \mathbf{0}$
            $\mathbf{w} \leftarrow \mathbf{0}$

Step 2   (iterative balancing)
            repeat
                for all measured links i
                    $\ln(\mathbf{h}) \leftarrow -g\alpha - \mathbf{A}''^{\mathrm{T}}\mathbf{u} + \mathbf{A}'^{\mathrm{T}}\mathbf{w}$
                    $w_i \leftarrow w_i + \ln(v_i') - \ln(\mathbf{a}_i'^{\mathrm{T}}\mathbf{h})$
                for all unmeasured links i
                    $\ln(\mathbf{h}) \leftarrow -g\alpha - \mathbf{A}''^{\mathrm{T}}\mathbf{u} + \mathbf{A}'^{\mathrm{T}}\mathbf{w}$
                    if $s_i < \mathbf{a}_i''^{\mathrm{T}}\mathbf{h}$ then $u_i \leftarrow u_i - \ln(s_i) + \ln(\mathbf{a}_i''^{\mathrm{T}}\mathbf{h})$
            until convergence.

In summary, the above log-linear programming problem yields the stochastic user equilibrium path flows, as well as the equilibrium link flows and delays for the capacity constrained links (with the provisos that every path considered is measured at least once and that the network has sufficient capacity to carry the trip table demands).

Consider the example in Fig. 7.1 when the link flow measurements, costs and capacities are as in Table 7.9. The flows, capacities and shadow prices are given in Table 7.15 for the dispersion parameter equal to 0.1. The symmetry of the solution follows from the symmetry of the problem. The corresponding fitted path flows are given in Table 7.16. Again, the symmetry of the assignment is evident.

Consider now the network in Fig. 7.3 and the flows, capacities and costs set out in Table 7.13. The link flows and shadow prices are listed in Table 7.17 for the case where the dispersion parameter is equal to 0.1. It is evident from the results that some over-adjustments must have occurred in early iterations, because delays arise on links which are not quite at capacity.

When Step 2 of the algorithm is modified to reduce the adjustment step size as follows:

Step 2 (iterative balancing)
    repeat
        for all measured links i
            $\ln(\mathbf{h}) \leftarrow -\mathbf{g}\alpha - \mathbf{A}''^T\mathbf{u} + \mathbf{A}'^T\mathbf{w}$
            $z_i \leftarrow z_i + \ln(v_i') - \ln(\mathbf{a}_i'^T\mathbf{h})$
        for all unmeasured links i
            $\ln(\mathbf{h}) \leftarrow -\mathbf{g}\alpha - \mathbf{A}''^T\mathbf{u} + \mathbf{A}'^T\mathbf{w}$
            if $s_i < \mathbf{a}_i''^T\mathbf{h}$ then $u_i \leftarrow u_i - \lambda(\ln(s_i) - \ln(\mathbf{a}_i''^T\mathbf{h}))$
    until convergence.

Table 7.15 *Flows, capacities and shadow prices*

Link	Capacity	Flow	Shadow price
1 (measured)		10	19.7183
2 (measured)		10	19.7183
3 (unmeasured)	3	3	6.4730
4 (unmeasured)	3	3	6.4730
5 (unmeasured)	10	3.7754	0
6 (unmeasured)	10	3.7754	0

Table 7.16 *Fitted path flows*

Path	1	2	3	4	5	6	7	8
Links	1	2,3,6	2	1,4,5	2,3	1,5	1,4	2,6
Flow	4.3572	1.1326	4.3572	1.1326	1.8674	2.6428	1.8674	2.6428

Table 7.17 *Flows, capacities and shadow prices*

Link	Capacity	Flow	Shadow price
1 (measured)		5	28.0595
2 (unmeasured)	5	3.0221	0
3 (measured)		5	28.0346
4 (unmeasured)	2	1.9779	13.7809
5 (unmeasured)	5	3.0221	0
6 (unmeasured)	5	3.0146	0
7 (unmeasured)	2	1.9633	6.4827
8 (unmeasured)	10	5.0221	0
9 (unmeasured)	5	3.0146	0
10 (unmeasured)	10	4.9779	0
11 (unmeasured)	2	1.9854	10.7180
12 (unmeasured)	2	2	3.2976

where $\lambda < 1$, the over-adjustment may be avoided. When $\lambda = 0.3$, the results given in Table 7.18 were obtained to three decimal places. Contrasting Tables 7.17 and 7.18, it would seem that the shadow prices have changed appreciably, but this merely reflects the indeterminacy in the shadow prices due to the inclusion of links with linearly dependent flows (in Chapter 3 it is shown that the dual variables are only unique if the constraints are linearly independent). The *path* shadow prices have not changed as significantly.

The corresponding fitted path flows are given in Table 7.19, again to three decimal places. As a result of the capacity constraints in the centre, most flow is forced to use the costlier periphery routes.

The results of the log-linear path flow estimator have been determined by the traffic flow measurements $\mathbf{v}'$, the active link capacity constraints $\mathbf{s}_a''$ and the path costs $\mathbf{g}$. It is therefore important to analyse the sensitivity of the fitted path flows and other outputs to possible error in these inputs. This may be done as before using the Tobin and Friesz (1988) method, provided all constraints are linearly independent.

At optimality, the following relationships hold for small changes in the primal and dual variables, $\Delta\mathbf{h}$, $\Delta\mathbf{u}_a$ and $\Delta\mathbf{w}$ respectively,

$$\Delta \mathbf{g}\alpha = \mathbf{D}\Delta\mathbf{h} + \mathbf{A}'^T\Delta\mathbf{w} - \mathbf{A}_a''^T\Delta\mathbf{u}_a$$

$$\Delta\mathbf{v}' = \mathbf{A}'\Delta\mathbf{h}$$

$$\Delta\mathbf{s}_a'' = \mathbf{A}_a''\Delta\mathbf{h}$$

where $\mathbf{D}$ is a diagonal matrix with elements $-1/h_i^*$ on the principal diagonal. Subscript a identifies the set of links with active capacity constraints. Small changes in the path costs, the flow measurements and the active capacity constraints are therefore related to small changes in the primal and dual variables as follows:

$$\begin{bmatrix} \Delta\mathbf{g} \\ \Delta\mathbf{v}' \\ \Delta\mathbf{s}_a'' \end{bmatrix} = \begin{bmatrix} \mathbf{D}/\alpha & \mathbf{A}^T/\alpha & -\mathbf{A}_a''^T/\alpha \\ \mathbf{A}' & \mathbf{0} & \mathbf{0} \\ \mathbf{A}_a'' & \mathbf{0} & \mathbf{0} \end{bmatrix} \begin{bmatrix} \Delta\mathbf{h} \\ \Delta\mathbf{w} \\ \Delta\mathbf{u}_a \end{bmatrix}$$

Table 7.18 *Flows, capacities and shadow prices*

Link	Capacity	Flow	Shadow price
1 (measured)		5	27.986
2 (unmeasured)	5	3	0
3 (measured)		5	27.986
4 (unmeasured)	2	2	10.755
5 (unmeasured)	5	3	0
6 (unmeasured)	5	3	0
7 (unmeasured)	2	2	9.232
8 (unmeasured)	10	5	0
9 (unmeasured)	5	3	0
10 (unmeasured)	10	5	0
11 (unmeasured)	2	2	7.756
12 (unmeasured)	2	2	6.232

Table 7.19 *Fitted path flows*

Path	1	2	3	4	5	6
Links	1,4,7,10	1,4,12,8	3,11,7,10	3,11,12,8	3,6,9,10	1,2,5,8
Flow	1	1	1	1	3	3

The matrix between the two perturbation vectors is the Jacobian of the system. The inverse of this Jacobian provides the required sensitivities, *provided the constraints included are linearly independent*:

$$\begin{bmatrix} \Delta h \\ \Delta w \\ \Delta u_a \end{bmatrix} = \begin{bmatrix} J_{11} & J_{12} & J_{13} \\ J_{21} & J_{22} & J_{23} \\ J_{31} & J_{32} & J_{33} \end{bmatrix} \begin{bmatrix} \Delta g \\ \Delta v \\ \Delta s_a \end{bmatrix}$$

Hence assuming that the active capacity constraints remain active, and that the inactive capacity constraints remain inactive, the perturbations have the following effect on equilibrium path flows:

$$\Delta h^* = J_{11} \Delta g + J_{12} \Delta v' + J_{13} s_a''$$

Variances and covariances may be obtained from the sensitivities. In particular, when path costs and link capacities remain unaltered,

$$\text{Var}\{h^*\} \approx J_{12} \text{Var}\{v'\} J_{12}^T$$

As an example, consider the network in Fig. 7.1 and the link flow measurements, costs and capacities given in Table 7.9. Table 7.20 gives the sensitivities of the fitted path flows to the link flow measurements and link capacities. The sensitivity of path flows to path costs is given in Table 7.21, noting that the dispersion parameter is 0.1.

Table 7.20 *Path flow sensitivities to link measurements and capacities*

		Measurement		Capacity	
Path	Links	Link 1	Link 2	Link 3	Link 4
1	1	0.622	0	0	−0.622
2	2,3,6	0	0	0.378	0
3	2	0	0.622	−0.622	0
4	1,4,5	0	0	0	0.378
5	2,3	0	0	0.622	0
6	1,5	0.378	0	0	−0.378
7	1,4	0	0	0	0.622
8	2,6	0	0.378	−0.378	0

Table 7.21 *Sensitivity of path flows to path costs*

Path flow	Path costs							
	1	2	3	4	5	6	7	8
1	−0.1645	0	0	0	0	0.1645	0	0
2	0	−0.0705	0	0	0.0705	0	0	0
3	0	0	−0.1645	0	0	0	0	0.1645
4	0	0	0	−0.0705	0	0	0.0705	0
5	0	0.0705	0	0	−0.0705	0	0	0
6	0.1645	0	0	0	0	0.1645	0	0
7	0	0	0	0.0705	0	0	−0.0705	0
8	0	0	0.1645	0	0	0	0	−0.1645

7.7 TIME-DEPENDENT METHODS

A number of authors have looked at the estimation of trip tables where the assignment is proportional but time-dependent (see Cascetta *et al.*, 1993, and Ashok and Ben-Akiva, 1993). This is a rather unsatisfactory compromise approach, as proportional assignment is generally not appropriate where congestion is a feature, but it is primarily in congested networks where time-dependent methods are most required.

In Chapter 6, it was suggested that a more realistic description of traffic assignment, particularly in congested networks, requires the introduction of the time dimension. One way of doing this is to build the time dimension into the network. As described in Chapter 3, nodes in a space–time extended network (STEN) constitute points not just in space but also in time. The links are assumed to have fixed capacities. When traffic cannot exit a link in the current period it is queued to leave the link in a later period. The queuing process is represented by links connecting a node to itself in a later period.

A difficulty with the first-in, first-out (FIFO) principle in connection with dynamic assignment has been identified in earlier chapters. FIFO leads to a non-convex set of feasible path flows (Carey, 1992). While queues obviously do exhibit a FIFO discipline, this is not so for traffic on links where overtaking is possible, so the desirability of a *rigid* enforcement of FIFO is in any case questionable.

In the case of stochastic user equilibrium assignment, where some sub-optimal behaviour is permitted, all feasible paths are used (although with decreasing probability as the cost of the path increases). In a STEN, some traffic would opt to remain in a queue when not actually required to do so by a capacity constraint. A model which allows the spreading of traffic across paths in the spatial domain but requires FIFO-like behaviour in the temporal domain has been described in Chapters 5 and 6. The same model may be used for path flow estimation.

Noting that queues in one period can be processed in subsequent periods, a potentially useful compromise is the approximation of a dynamic assignment by a sequence of steady state equilibria linked across time by queues. Equilibrium delay is a penalty required to bring demand into line with a fixed supply of infrastructure. However, underlying the notion of delay is some form of queuing, so a natural extension would be to translate equilibrium delays into equilibrium queues which may be carried between periods. By explicitly introducing queues, the temporary overloading of links can be modelled.

The equilibrium queue is equal to the equilibrium delay times the service rate. Let S be a diagonal matrix of link service rates

$$\mathbf{S} = \begin{bmatrix} s_1 & & \\ & s_2 & \\ & & \ddots \end{bmatrix}$$

then

$$\mathbf{d}_t = \mathbf{S}^{-1}\mathbf{q}_t$$

where $\mathbf{d}_t$ is a vector of equilibrium link delays in period t and $\mathbf{q}_t$ is a vector of equilibrium link queues in period t.

As previously, the links are partitioned into two sets, the first set being those links for which a flow measurement is made each period and the second set being those for which this is not the case. For both sets of links, the link cost and the link capacity are known. The link-path incidence matrix is accordingly partitioned into $\mathbf{A}'$ for the measured links and $\mathbf{A}''$ for the unmeasured links. For the measured links, $\mathbf{v}_t'$ is known while for the unmeasured links, only the capacity is known. It is assumed that points at which the flow is measured (in practice, the vehicle detectors) are located upstream, near the entrance of the link, so that the *demand* is measured free of any disturbances due to queuing. This assumed positioning of the vehicle detectors is consistent with many urban traffic control systems, in particular SCOOT (see Hunt *et al.*, 1982).

The link capacity constraints have the form of the following complementary slackness conditions

$$\mathbf{s} - \mathbf{Ah}_t - \mathbf{q}_{t-1} + \mathbf{q}_t \geqslant \mathbf{0}$$

$$\mathbf{q}_t \geqslant \mathbf{0}$$

$$(\mathbf{s} - \mathbf{Ah}_t - \mathbf{q}_{t-1} + \mathbf{q}_t)^{\mathrm{T}}\mathbf{q}_t = 0$$

where $\mathbf{q}_{t-1}$ is the (given) vector of queues at the end of period $t-1$ and $\mathbf{q}_t$ is the (to be estimated) vector of queues at the end of period t. The demand for the links $\mathbf{Ah}_t$ can therefore exceed their steady state capacities $\mathbf{s}$ if the equilibrium queues are growing between periods, namely if $\mathbf{q}_t - \mathbf{q}_{t-1} > \mathbf{0}$. The complementary slackness conditions imply that the queue carried over from the preceding period is processed before the new arrivals, and that a new queue is only formed when the demand exceeds the available capacity. Although the composition of the queue is not considered, the macroscopic effects of FIFO are enforced.

The model presented here is related to the model proposed by Lam *et al.* (1995). The following objective function

$$f(\mathbf{h}_t, \mathbf{q}_t) = (\mathbf{h}_t^{\mathrm{T}}(\ln(\mathbf{h}_t) - \mathbf{1}) + \alpha\mathbf{h}_t^{\mathrm{T}}\mathbf{g} + 0.5\alpha\mathbf{q}_t^{\mathrm{T}}\mathbf{S}^{-1}\mathbf{q}_t$$

where $\mathbf{g}$ is the vector of path costs, is minimised subject to link flow measurements

$$\mathbf{v}_t' = \mathbf{A}'\mathbf{h}_t$$

and link capacity constraints

$$\mathbf{s} \geqslant \mathbf{Ah}_t + \mathbf{q}_{t-1} - \mathbf{q}_t$$

The Hessian of $f(\mathbf{h}_t, \mathbf{q}_t)$ is positive definite, so the objective function is strictly convex. The constraints are also convex, so the solution to this minimisation problem is unique in the primal variables $\mathbf{h}_t$ and $\mathbf{q}_t$. The Lagrangian equation has the form

$$L(\mathbf{h}_t, \mathbf{q}_t, \mathbf{u}_t, \mathbf{w}_t) = f(\mathbf{h}_t, \mathbf{q}_t) + \mathbf{w}_t^{\mathrm{T}}(\mathbf{v}_t' - \mathbf{A}'\mathbf{h}_t) - \mathbf{u}_t^{\mathrm{T}}(\mathbf{s} - \mathbf{Ah}_t - \mathbf{q}_{t-1} + \mathbf{q}_t)$$

The optimality conditions for $\mathbf{q}_t$, namely

$$\nabla_q L(\mathbf{h}_t, \mathbf{q}_t^*, \mathbf{u}_t, \mathbf{w}_t) = \mathbf{S}^{-1}\mathbf{q}_t^*\alpha - \mathbf{u}_t = \mathbf{0},$$

identify the dual variables $\mathbf{u}_t$ as being equal to the equilibrium delays multiplied by α, since

$$\mathbf{S}^{-1}\mathbf{q}_t\alpha = \mathbf{u}_t = \mathbf{d}_t\alpha$$

The optimality conditions for $\mathbf{h}_t$ are

$$\ln(\mathbf{h}_t^*) = -\mathbf{g}_t\alpha - \mathbf{A}^{\mathrm{T}}\mathbf{u}_t + \mathbf{A}'^{\mathrm{T}}\mathbf{w}_t = -\mathbf{g}_t\alpha - \mathbf{A}'^{\mathrm{T}}(\mathbf{u}_t' - \mathbf{w}_t) - \mathbf{A}''^{\mathrm{T}}\mathbf{u}_t''$$

where $\mathbf{w}_t$ may be interpreted as the vector of shadow prices for the marginal trip on each measured link. Note that for measured links, consistency requires that the capacity constraint is inactive, so $\mathbf{u}_t'$ is zero for all elements. Hence

$$\ln(\mathbf{h}_t^*) = -\mathbf{g}_t\alpha + \mathbf{A}'^T\mathbf{w}_t - \mathbf{A}''^T\mathbf{u}_t''$$

As explained in the previous section, this may be regarded as the logit path choice model. The optimality conditions for $\mathbf{u}_t$ are

$$\mathbf{s} - \mathbf{A}\mathbf{h}_t - \mathbf{q}_{t-1} + \mathbf{q}_t \geqslant \mathbf{0}$$

$$\mathbf{u}_t \geqslant \mathbf{0}$$

$$\mathbf{u}_t^T(\mathbf{s} - \mathbf{A}\mathbf{h}_t - \mathbf{q}_{t-1} + \mathbf{q}_t) = \mathbf{0}$$

Noting that a positive element of $\mathbf{u}_t$ corresponds to a positive element of $\mathbf{q}_t$ and a zero element of $\mathbf{u}_t$ corresponds to a zero element of $\mathbf{q}_t$, these optimality conditions give the required capacity–queue complimentary slackness conditions given earlier.

An examination of the objective function reveals that, while trips are *encouraged* to spread spatially across paths by the entropy-related term $(\mathbf{h}_t^T(\ln(\mathbf{h}_t) - \mathbf{1})$, queuing is *discouraged* by the quadratic term $0.5\alpha\mathbf{q}_t^T\mathbf{S}^{-1}\mathbf{q}_t$. Although temporary overloading of the network can be accommodated through queue formation, FIFO is not an issue. Since the role of the queue in one period is to reduce the capacity of the link in the subsequent period, the composition of the queue in terms of user class, origin or destination is not relevant here.

This model may be fitted by either the method of successive averages or a modified version of the iterative balancing algorithm.

Method of Successive Averages for logit assignment

Step 1   (initialisation)
$$\mathbf{u}_t \leftarrow \mathbf{0}$$
$$\mathbf{w}_t \leftarrow \mathbf{0}$$
$$\mathbf{q}_t \leftarrow \mathbf{0}$$
$$n \leftarrow 1$$

Step 2   (successive averaging)
repeat
    for all measured links i
        $\ln(\mathbf{h}_t) \leftarrow -\alpha\mathbf{A}^T\mathbf{c} - \mathbf{A}^T\mathbf{u}_t + \mathbf{A}'^T\mathbf{w}_t$
        if $v_{i,t}' \neq \mathbf{a}_i'^T\mathbf{h}_t$ then
        $w_{i,t} \leftarrow w_{i,t} + \ln(v_{i,t}') - \ln(\mathbf{a}_i'^T\mathbf{h}_t)$
    for all links i
        $\ln(\mathbf{h}_t) \leftarrow -\alpha\mathbf{A}^T\mathbf{c} - \mathbf{A}^T\mathbf{u}_t + \mathbf{A}'^T\mathbf{w}_t$
        $v_{i,t} \leftarrow \sum_{\text{all}j} a_{ij}h_{j,t}$

        if $s_i < v_{i,t} + q_{i,t-1}$ $\begin{array}{l}\text{then } q_{i,t}^* \leftarrow v_{i,t} + q_{i,t-1} - s_i \\ \text{else } q_{i,t}^* \leftarrow 0\end{array}$
        $q_{i,t} \leftarrow (1/n)q_{i,t}^* + (1 - 1/n)q_{i,t}$
        $u_{i,t} \leftarrow q_{i,t}/s_i$
    $n \leftarrow n + 1$
until convergence.

In Step 2 of this simple but effective algorithm, the $\mathbf{w}$ variables are first calculated so that the measured link flows $\mathbf{v}'$ are reproduced. Then for each link i, a comparison of the current demand with the capacity leads to the definition of an auxiliary queue. This is then averaged with the current queue and the corresponding link delay is determined.

The application of iterative balancing is not so simple in this case.

Iterative balancing algorithm for logit assignment

Step 1   (initialisation)
$$\mathbf{w}_t \leftarrow \mathbf{0}$$
$$\mathbf{q}_t \leftarrow \text{Max}\{\mathbf{0}, \mathbf{q}_{t-1} - \mathbf{s}_t + \mathbf{0.0001}\}$$
$$\mathbf{u}_t \leftarrow \mathbf{S}^{-1}\mathbf{q}_t\alpha$$

Step 2   (iterative balancing)
repeat
   for all measured links i
   $$\ln(\mathbf{h}_t) \leftarrow -\alpha\mathbf{A}^T\mathbf{c} - \mathbf{A}^T\mathbf{u}_t + \mathbf{A}'^T\mathbf{w}_t$$
   if $v_{i,t}' \leftarrow \mathbf{a}_i'^T\mathbf{h}_t$ then
   $$w_{i,t} \leftarrow w_{i,t} + \ln(v_{i,t}') - \ln(\mathbf{a}_i'^T\mathbf{h}_t)$$
   for all links i
   $$\ln(\mathbf{h}_t) - \alpha\mathbf{A}^T\mathbf{c} - \mathbf{A}^T\mathbf{u}_t + \mathbf{A}'^T\mathbf{w}_t$$
   if $s_i + q_{i,t} - q_{i,t-1} < \mathbf{a}_i^T\mathbf{h}_t$ then
   $$m_i \leftarrow (1 + s_i/\alpha(s_i - q_{i,t-1} + q_{i,t}))$$
   $$u_{i,t} \leftarrow u_{i,t} - (\ln(s_i - q_{i,t-1} + q_{i,t}) - \ln(\mathbf{a}_i^T\mathbf{h}_t))/m_i$$
   $$q_{i,t} \leftarrow s_i u_{i,t}/\alpha$$
until convergence.

As in the method of successive averages, Step 2 starts by calculating the **w** variables so that the measured link flows are reproduced. If the current as well as the past queue on link i were *fixed*, then the adjustment

$$u_{i,t} \leftarrow u_{i,t} + \text{Max}\{-\ln(s_i - q_{i,t-1} + q_{i,t}) + \ln(\mathbf{a}_i^T\mathbf{h}_t), 0\}$$

(made only when the current demand plus the past queue exceeds the capacity plus the current queue) would lead to a convergent algorithm, because as shown in Chapter 3 the direction of change is a ascent direction and the size of the change is not sufficient to take the dual variable $u_{i,t}$ beyond its optimal value (the *ascent* does not become a *descent*). However, in this case

$$q_{i,t} = s_i u_{i,t}/\alpha$$

so the ascent could well become a descent if the above adjustment were made. As shown in Chapter 6, the scaling down of the adjustment by the following factor:

$$m_i = (1 + s_i/\alpha(s_i - q_{i,t-1} + q_{i,t}))$$

prevents over-adjustment.

When a sequence of time periods is considered, the above algorithm may be applied recursively. Starting with $\mathbf{q}_0$, $\mathbf{q}_1$ may be estimated, then $\mathbf{q}_2$ may be estimated from $\mathbf{q}_1$, etc. This is illustrated in the example below.

The critical factors in this path flow estimator are the link flow measurements, the link capacities and the path costs. One would therefore be interested in the sensitivity of the fitted path flows to these parameters. The Tobin and Friesz (1988) approach to sensitivities may be applied here. At the optimum, the following relationships hold for small changes in the primal variables, $\Delta\mathbf{h}$ and $\Delta\mathbf{q}_a$, and the dual variables, $\Delta\mathbf{w}$

$$
\begin{bmatrix} \Delta\mathbf{g} \\ \Delta\mathbf{v}_t' \\ \Delta\mathbf{s}_a \end{bmatrix} = \begin{bmatrix} \mathbf{D}_t/\alpha & \mathbf{A}'^T/\alpha & -\mathbf{A}_a^T\mathbf{S}_a^{-1} \\ \mathbf{A}' & 0 & 0 \\ \mathbf{A}_a & 0 & -\mathbf{I} \end{bmatrix} \begin{bmatrix} \Delta\mathbf{h}_t \\ \Delta\mathbf{w}_t \\ \Delta\mathbf{q}_{a,t} \end{bmatrix}
$$

where $\mathbf{D}_t$ is a diagonal matrix with elements $-1/h_{i,t}^*$ on the principal diagonal. Subscript a identifies the set of links with active capacity constraints. The matrix between the two perturbation vectors is the Jacobian of the system. The inverse of this Jacobian provides the required sensitivities, *provided the constraints included are linearly independent*. The inverse of this Jacobian, represented in block form below, gives the sensitivities.

$$
\begin{bmatrix} \Delta\mathbf{h}_t \\ \Delta\mathbf{w} \\ \Delta\mathbf{q}_a \end{bmatrix} = \begin{bmatrix} \mathbf{J}_{11} & \mathbf{J}_{12} & \mathbf{J}_{13} \\ \mathbf{J}_{21} & \mathbf{J}_{22} & \mathbf{J}_{23} \\ \mathbf{J}_{31} & \mathbf{J}_{32} & \mathbf{J}_{33} \end{bmatrix} \begin{bmatrix} \Delta\mathbf{g} \\ \Delta\mathbf{v}' \\ \Delta\mathbf{s}_a \end{bmatrix}
$$

Hence assuming that the active capacity constraints remain active and the inactive capacity constraints remain inactive, the perturbations have the following effect:

$$
\Delta\mathbf{h}_t^* = \mathbf{J}_{11}\Delta\mathbf{g} + \mathbf{J}_{12}\Delta\mathbf{v}_t' + \mathbf{J}_{13}\Delta\mathbf{s}_a
$$

Variances and covariances may be obtained from the sensitivities. In particular, when path costs and link capacities remain unaltered,

$$
\text{Var}\{\mathbf{h}_t^*\} \approx \mathbf{J}_{12}\text{Var}\{\mathbf{v}_t'\}\mathbf{J}_{12}^T
$$

Returning to the network portrayed in Fig. 7.1, the build up and decline of queues over the peak period is simulated by inputting the link measurements given in Table 7.22. As mentioned earlier, it is assumed that the detectors are located near the entrance of the links, so that demand is measured (in say vehicles per minute). The value of the dispersion parameter α was taken to be 0.1.

The link flows and capacities are shown in Table 7.23. The link flows here are calculated from the path flows and represent the demand for the link, which is allowed to exceed the capacity. The capacities of links 1 to 4 are exceeded in the second period. Demand falls below capacity on links 1 and 2 (which cannot be avoided) and links 3 and 4 (which can be avoided) in period 6.

The evolution of the queues is shown in Table 7.24. The queue falls to zero on links 1 and 2 but persists on links 3 and 4, due to the 'suppressed demand' from alternative higher cost paths not using links 3 and 4. This would be one example of induced traffic.

The sensitivities to path cost for period eight are shown in Table 7.25. As a result of the trip table total constraints, the effect of path cost changes is limited to the element of the trip table to which the path belongs. Note that the row and column sums of the sensitivities are zero.

Table 7.22 *Link measurements over eight periods*

	Link demand by period [units of flow]							
Link	1	2	3	4	5	6	7	8
1	5	10	11	10	7	5	4	3
2	5	10	11	10	7	5	4	3

Table 7.23 *Link capacities and demands*

		Link demand by period [units of flow]							
Link	Capacity	1	2	3	4	5	6	7	8
1	8	5	10	11	10	7	5	4	3
2	8	5	10	11	10	7	5	4	3
3	3	2.251	4.388	4.676	4.157	2.914	2.118	1.735	1.342
4	3	2.251	4.388	4.676	4.157	2.914	2.118	1.735	1.342
5	10	1.888	3.775	4.153	3.775	2.643	1.888	1.510	1.133
6	10	1.888	3.775	4.153	3.775	2.643	1.888	1.510	1.133

Table 7.24 *Link costs and queues*

		Link queue by period [units of flow x period]							
Link	Cost	1	2	3	4	5	6	7	8
1	5	0	2	5	7	6	3	0	0
2	5	0	2	5	7	6	3	0	0
3	2	0	1.388	3.063	4.220	4.134	3.251	1.987	0.329
4	2	0	1.388	3.063	4.220	4.134	3.251	1.987	0.329
5	5	0	0	0	0	0	0	0	0
6	5	0	0	0	0	0	0	0	0

Table 7.25 *Sensitivity of path flow to path cost in period eight*

		Sensitivity to change of cost on path [cost units]							
Path	Links	1	2	3	4	5	6	7	8
1	1	−0.067	0	0	0.017	0	0.022	0.028	0.028
2	2,3,6	0	−0.042	0.017	0	0.015	0	0	0
3	2	0	0.017	−0.067	0	0.028	0	0	0
4	1,4,5	0.017	0	0	−0.042	0	0.010	0.015	0.015
5	2,3	0	0.015	0.028	0	−0.060	0	0	0
6	1,5	0.022	0	0	0.010	0	−0.049	0.017	0.017
7	1,4	0.028	0	0	0.015	0	0.017	−0.060	−0.060
8	2,6	0	0.010	0.022	0	0.017	0	0	0

The sensitivities of path flows to the measurements on links 1 and 2 and the capacities of links 3 and 4 in the eighth period are shown in Table 7.26. Note that the sensitivities sum to one for each measured link, indicating that flow is conserved (every measured unit of flow appears on some path). The sensitivities sum to zero for each constrained link.

The time-dependent methods presented so far have been based on deterministic queuing relationships of the form

$$q_t = Max\{Ah_t + q_{t-1} - s, 0\}$$

where the maximum operator is applied to each pair of elements. This leaves out of account both the random component of delay as well as the uniform component of delay under traffic signal control. Kimber and Hollis (1979) have presented more realistic time-dependent delay expressions that relate in principle to any kind of junction (see Chapter 4). These can be conveniently incorporated into the method of successive averages as follows.

Table 7.26 *Sensitivity of path flow to link demand and link capacity*

Path	Links	Measurement		Constraint	
		Link 1	Link 2	Link 3	Link 4
1	1	0.351	0	0	−0.015
2	2,3,6	0	0.165	0.009	0
3	2	0	0.351	−0.015	0
4	1,4,5	0.165	0	0	0.009
5	2,3	0	0.272	0.015	0
6	1,5	0.213	0	0	−0.009
7	1,4	0.272	0	0	0.015
8	2,6	0	0.213	−0.009	0

Method of Successive Averages for logit assignment with K—H delays

Step 1 (initialisation)

$\mathbf{u}_t \leftarrow \mathbf{0}$

$\mathbf{w}_t \leftarrow \mathbf{0}$

$\mathbf{q}_t \leftarrow \mathbf{0}$

$n \leftarrow 1$

Step 2 (successive averaging)

repeat

for all measured links i

$\ln(\mathbf{h}_t) \leftarrow -\alpha \mathbf{A}^T\mathbf{c} - \mathbf{A}^T\mathbf{u}_t + \mathbf{A}'^T\mathbf{w}_t$

if $v_{i,t}' \neq \mathbf{a}_i'^T\mathbf{h}_t$ then

$w_{i,t} \leftarrow w_{i,t} + \ln(v_{i,t}') - \ln(\mathbf{a}_i'^T\mathbf{h}_t)$

for all links i

$\ln(\mathbf{h}_t)/\alpha \leftarrow -\mathbf{A}^T\mathbf{c} - \mathbf{A}^T\mathbf{u}_t + \mathbf{A}'^T\mathbf{w}_t$

$v_{i,t} \leftarrow \sum_{\text{all } j} a_{ij}h_{j,t}$

$q_{i,t}, u_{i,t}^* \leftarrow$ Kimber and Hollis (1979) $\leftarrow v_{i,t}, s_{i,t}, q_{i,t-1}$

$u_{i,t} \leftarrow (1/n)u_{i,t}^* + (1 - 1/n)u_{i,t}$

$n \leftarrow n + 1$

until convergence.

While deterministic queuing relationships are probably the most convenient to use when little is known about the links, the Kimber and Hollis (1979) expressions should yield more accurate path flow estimates.

7.8 CONCLUSIONS

This chapter has reviewed the salient approaches to the estimation of trip tables. Particular attention has been given to the derivation of sensitivity information, so that users are in a position to assess the robustness of results to errors in the inputs. Methods based on proportional assignment are gradually giving way to methods based on deterministic user equilibrium assignment, which are inherently non-proportional unless link costs are fixed. Bi-level programming offers an effective approach to the estimation of trip tables, when the Tobin and Friesz (1988) deterministic user equilibrium sensitivity expressions are used in the manner suggested by Yang (1995).

Transportation planners and traffic engineers are generally interested in path-based information, rather than the more aggregated trip table *per se*. The use of stochastic user equilibrium assignment not only leads to greater realism but overcomes the non-identifiability of path flows under deterministic user equilibrium assignment. This chapter introduces a stochastic user equilibrium path flow estimator.

Increasingly, attention is being given to approaches to trip table estimation that allow trips to span a number of time periods. A time-dependent path flow estimator based on stochastic user equilibrium assignment is presented which avoids the difficulties of non-convexity generally associated with dynamic assignment. Equilibrium is assumed to prevail in each period, but periods are linked by queues (a minor inconsistency in assumptions, as equilibrium presupposes that all trips are completed while queues constitute incomplete trips). The macroscopic effects of FIFO are respected, because precedence is given to existing queues over new arrivals.

8
Network Reliability

8.1 INTRODUCTION

Whether or not the capacity of the transportation network is satisfactory can be evaluated by various measures, such as the travel times and the extent of congestion. In the absence of direct observations, estimates of these measures may be obtained from a *traffic assignment model* (dealt with in Chapters 5 and 6). However, fluctuations in demand are generally not considered in traffic assignment models. In addition, network flows are influenced by *abnormal events* that affect network characteristics and capacity, like disasters, accidents, construction or repair. Ideally, networks should be designed so as to cope with normal fluctuations by offering alternative paths, but planning for abnormal events is much more difficult.

In *systems engineering*, reliability may be defined as the degree of stability of the quality of service which a system *normally* offers. In the face of increasing user demands for high levels of service, system reliability is becoming increasingly important in the planning, construction and operation of transportation networks.

In evaluating network reliability, the flow (of people, vehicle, goods, etc.) may be divided into *normal* and *abnormal* states. Moreover, transportation network reliability has two aspects; *connectivity* and *travel time* reliability. Connectivity is the probability that traffic can reach a given destination at all, while travel time reliability is the probability that traffic can reach a given destination within a given time. This chapter concentrates on network connectivity in normal situations. For travel time reliability, concepts only will be touched on, since studies on this subject are scarce. Regarding network reliability in abnormal situations, many issues remain to be solved.

Transportation networks in urban and rural areas can be large and have complicated topologies, so reliability analysis can necessitate a large amount of computation. Reliability analysis for transportation networks differs from general system reliability analysis, in that path choice behaviour must be considered. In communication networks, for example, shifting from a shorter path to a much longer path when the shorter path becomes unavailable usually poses no problem, whereas in transportation networks, tripmakers are assumed to prefer shorter, faster or, more generally, less costly paths. A review of the existing literature relating to reliability analysis in systems engineering and transportation networks is to be found in Du and Nicholson (1993).

8.2 CONNECTIVITY

Equipment comprising many interrelated components is generically called a *system*. Each component or part of a system is called a *unit*. In systems engineering, reliability has been defined as 'the ability of an item to perform a required function, under stated conditions, for a stated period of time' (see Pages and Gondran, 1986). Units are presumed to have two states; *function* and *failure*. The reliability of the unit is the probability that it is operational during a specified period.

The *reliability graph* expresses the linkages between the *reliability* of each unit in a system of interconnected units. Analysis of the graph allows the degree of connectivity from the input node to the output node of the system to be determined (Inoue, 1976, and Henley and Kumamoto, 1981). Because of its conceptual and arithmetic simplicity, this method of analysis is often used for reliability analysis. In the case of transportation networks, a link in the reliability graph corresponds to a physical link (a road or railway line, for example). Thus reliability for a given origin–destination pair (referred to as a *terminal reliability*) can be represented by the degree of connectivity between the corresponding two nodes (or centroids) in the reliability graph.

The reliability of the links can be determined in various ways. For simplicity, consider the case where link reliability corresponds to the probability that the corresponding physical link congests, which occurs when the link flow is greater than the link capacity. If 50 out of 100 measurements of hourly flow on a link are greater than its downstream capacity, the link connectivity or reliability may be 0.5. Given the link reliabilities, the connectivity between two points (perhaps an origin and a destination) can be calculated by reliability analysis, as presented below. If this connectivity were, say, 0.6, the flow from the specified origin can reach a specified destination without encountering congestion 6 times out of 10. This applies to *normal conditions* only, not to abnormal situations such as disasters, accidents, and construction work. In abnormal situations, a different approach must be taken.

Link reliability can also be defined in terms of travel time. In this case, transportation network reliability could be the 'probability that trips between from a given origin can reach a given destination within a given period of time during stated hours'. This travel time reliability measure will be explained in Section 8.7. The following section describes the basic idea of transportation network reliability in terms of connectivity.

8.3   STRUCTURE FUNCTION

First, the state of congestion or otherwise of link i is represented by the 0–1 state variable x_i as follows

$$x_i = \begin{cases} 1 & \text{if link } i \text{ functions} \\ 0 & \text{otherwise} \end{cases}$$

With vector $\mathbf{x}$ as the state vector of the system, the condition of the system can be expressed as

$$\phi(\mathbf{x}) = \begin{cases} 1 & \text{if the system functions} \\ 0 & \text{if the system fails} \end{cases}$$

This function, $\phi(\mathbf{x})$, is called the *structure function*. The function is used to find the value of system reliability. When a system under consideration consists of a series system or a parallel system, as shown in Figs. 8.1 and 8.2 respectively, the structure function can be

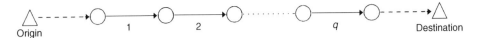

Figure 8.1   *A series system*

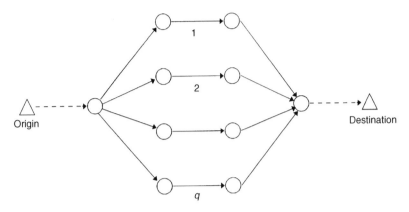

Figure 8.2   *A parallel system*

expressed by the following equations.

Series system: $\phi(\mathbf{x}) = \Pi_i x_i$

Parallel system: $\phi(\mathbf{x}) = 1 - \Pi_i(1 - x_i)$

When the system is a combination of series and parallel sub-systems, the structure function can be expressed as a combination of the above equations. In complicated systems, *decomposition* or *path-and-cut* methods are used to derive the structure function. The decomposition method expresses a structure function of order q as a structure function of order $(q - 1)$ (see for example, Shogan, 1978, or Nakazawa, 1981), which is possible provided the system can be decomposed into series and parallel structures. Such a decomposition is not always possible, and even when it is, the processing is painstaking and not always practical. The path-and-cut method, in contrast, is more practical, as system reliability can be calculated directly from the path a cut sets for the corresponding reliability graph.

A transportation system can be represented as a network consisting of a set of nodes plus centroids and a set of links (see Chapter 2). When at least one path exists between all pairs of nodes, the graph is said to be *connected*. The minimum number of successive links needed to connect a pair of nodes is called the *minimal path*. The minimum number of links needed to disconnect a pair of nodes is called the *cut set* (or minimal cut set). Link sets constituting paths between nodes i and j in the network in Fig. 8.3 include

$$(1, 2), (3, 4), (1, 5, 4)$$

Link sets constituting a cut set are

$$(1, 3), (2, 4), (1, 4), (2, 5, 3)$$

Following Mine and Kawai (1982), the system of Fig. 8.3 may be viewed either as a *series system in parallel*, by using the paths (see Fig. 8.4), or as a *parallel system in series*, by using the cut sets (see Fig. 8.5). If any one of the three paths in Fig. 8.4 functions, the system functions. Likewise, if any one of the four cut sets in Fig. 8.5 fails, the system cannot function. The expression of the function/failure state of a system with the aid of series and parallel sub-systems is called the *equivalent transformation* of the system.

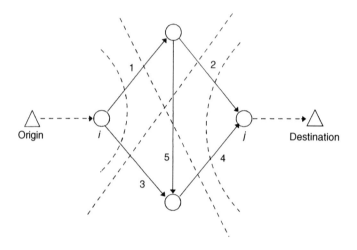

Figure 8.3   *Example network and cut sets*

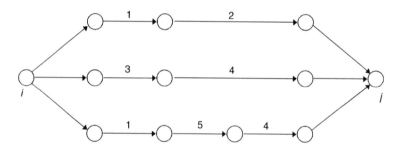

Figure 8.4   *Series systems in parallel*

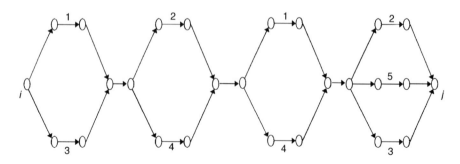

Figure 8.5   *Parallel systems in series*

Let the total number of paths and cut sets be p and k respectively, expressing the path sets as $P(1), P(2), \ldots, P(p)$ and the cut sets as $K(1), K(2), \ldots, K(k)$. Since the structure function for path s corresponds to a series system, it may be expressed as

$$\alpha_s(\mathbf{x}) = \prod_{i \in P(s)} x_i$$

Because the system consists of parallel paths $P(1), P(2), \ldots, P(p)$, the system structure function can be expressed as

$$\phi(\mathbf{x}) = 1 - \prod_{s=1 \text{ to } p} (1 - \alpha_s(\mathbf{x})) = 1 - \prod_{s=1 \text{ to } p} \left(1 - \prod_{i \in P(s)} x_i \right)$$

The structure function for the example in Fig. 8.3 expressed in terms of paths is

$$\phi(\mathbf{x}) = 1 - (1 - x_1 x_2)(1 - x_3 x_4)(1 - x_1 x_5 x_4)$$

Likewise, since the structure function $\beta_s(\mathbf{x})$ of cut set K_s is a parallel system, it can be expressed as

$$\beta_s = 1 - \prod_{i \in K(s)} (1 - x_i)$$

Because the system consists of a series of cut sets $K(1), K(2), \ldots, K(k)$, the system structure function may be expressed as

$$\phi(\mathbf{x}) = \prod_{s=1 \text{ to } k} \beta_s(\mathbf{x}) = \prod_{s=1 \text{ to } k} \left(1 - \prod_{i \in K(s)} (1 - x_i) \right)$$

In the case of Fig. 8.3, the equation is

$$\phi(\mathbf{x}) = (1 - (1 - x_1)(1 - x_3))(1 - (1 - x_2)(1 - x_4))(1 - (1 - x_1)(1 - x_4))$$
$$(1 - (1 - x_2)(1 - x_5)(1 - x_3))$$

When links 1 and 2 function correctly, and when links 3, 4 and 5 do not function at all, the link state vector $\mathbf{x}^T$ is (1, 1, 0, 0, 0). In this case, whichever structure function is used, the system remains connected, since $\phi(\mathbf{x}) = 1$, which confirms that the system is functioning. When links 1, 3 and 5 function and when links 2 and 4 do not, however, the link state vector is (1, 0, 1, 0, 1) and the system is disconnected since $\phi(\mathbf{x}) = 0$, which indicates that the system is not functioning.

8.4 STRUCTURE FUNCTION AND RELIABILITY VALUE

To find the reliability of a transportation system, the reliability of the links involved must first be determined. The link reliability value r_i is the expected value if x_i, which is assumed to be a random binary variable.

$$r_i = E\{x_i\}$$

When link flow can be measured, link reliability could be the frequency that the link flow is less than or equal to the link capacity. When it is not possible to measure link flow directly, an alternative would be simulation based on a traffic assignment model (Wakabayashi *et al.*, 1993).

The system reliability is the expected value of the structure function, namely

$$R = E\{\phi(\mathbf{x})\}$$

Thus the system reliability for a pair of nodes is as follows:

$$R = E \left\{ 1 - \prod_{s=1 \text{ to } p} \left(1 - \prod_{i \in P(s)} x_i \right) \right\}$$

for the path set based expression, or

$$R = E \left\{ \prod_{s=1 \text{ to } k} \left(1 - \prod_{i \in K(s)} (1 - x_i) \right) \right\}$$

for the cut set based expression.

For the example in Fig. 8.3, the path set based expression is

$$R = E\{1 - (1 - x_1 x_2)(1 - x_3 x_4)(1 - x_1 x_5 x_4)\}$$

and the cut set based expression is

$$R = E\{(1 - (1 - x_1)(1 - x_3))(1 - (1 - x_2)(1 - x_4))(1 - (1 - x_1)(1 - x_4))$$
$$(1 - (1 - x_2)(1 - x_5)(1 - x_3))\}$$

As the state variables of any link appear more than once in these expressions, Boolean algebra is required to evaluate R, whether the path or the cut set based expression is used.

When the system is only a series or a parallel system, the system reliability can be expressed as follows:

$$R = E \left\{ \prod_{i \in S} x_i \right\} = \prod_{i \in S} r_i$$

for a series system S, or

$$R = E \left\{ 1 - \prod_{i \in P} (1 - x_i) \right\} = 1 - \prod_{i \in P} (1 - r_i)$$

for a parallel system P.

Boolean algebra may require a tremendous amount of calculation, since the computation time and the memory requirement both increase exponentially with the number of links. Moreover, to obtain the exact reliability by this method, all paths, or alternatively all cut sets, between any pair of nodes must be identified. Although that may not pose any problem in communication networks, *some* paths are unrealistic for transportation networks (for example, paths involving cycles). As for cut sets, it is more practical to handle cross sections corresponding to screen lines or cordons. Thus, for transportation networks, the objective is to develop an efficient heuristic for the estimation of reliability. A number of exact methods are presented in Section 8.5, followed by a number of heuristics in Section 8.6.

8.5 EXACT METHODS

8.5.1 Combination method

When the system can be expressed as a combination of series and parallel systems, decomposition allows the direct calculation of system reliability. However, when the same link appears two or more times, as shown in Figs. 8.4 and 8.5, the combination method is not applicable. In this case, Boolean algebra is necessary, which makes the calculation complicated.

8.5.2 Enumeration method

This is based on the following decomposition:

$$R(\mathbf{r}) = \ \Pr\{x_i = 1\}R(1_i, \mathbf{r}') + \Pr\{x_i = 0\}R(0_i, \mathbf{r}')$$

where R $(1_i, \mathbf{r}')$ is the system reliability when unit i is functioning, R $(0_i, \mathbf{r}')$ the system reliability value when unit i is in failure, and $\mathbf{r}'$ is the state vector of link reliabilities *excluding* link i. If the system can be decomposed into lower-dimension systems that are series or parallel systems, or a combination of them, then by repeating the application of this equation, the system reliability can be calculated. This method is called the enumeration method. In complicated systems, however, it is difficult to achieve an equivalent transformation mechanically into lower-dimension system structures. If achieved, it needs a tremendous amount of calculation.

8.5.3 Inclusion–exclusion formula

Let the event in which all components in path $P(s)$ are functioning be E_s. If at least one path functions, the entire system functions. Therefore, the system reliability value is given by

$$R = \Pr\{\cup_{s=1\,to\,p}E_s\}$$

This method is the same as the method to find the expected value of a logical summation given by Venn diagrams (Inoue, 1976). Using the inclusion–exclusion formula, the reliability becomes

$$R = \sum_{s=1\,to\,p} \Pr\{E_s\} - \sum_{s=1\,to\,p} \sum_{all\,t\neq s} \Pr\{E_s \cap E_t\}$$

$$+ \sum_{s=1\,to\,p} \sum_{t\neq s} \sum_{u\neq s,t} \Pr\{E_s \cap E_t \cap E_u\} + \ldots + (-1)^{p-1}\ \Pr\{\cap_{s=1-p}E_s\}$$

In the case of cut sets, let the event in which all components in cut set $K(s)$ fail be $\underline{E}_s$ (the complement of E_s). If at least one cut set is generated, the system fails. Therefore, the probability F that the system fails is given by

$$F = \Pr\{\cup_{s=1\,to\,k}\underline{E}(s)\}$$

Following the same procedures,

$$F = \sum_{s=1 \text{ to } k} \Pr\{\underline{E}_s\} - \sum_{s=1 \text{ to } k} \sum_{\text{all } t \neq s} \Pr\{\underline{E}_s \cap \underline{E}_t\}$$

$$+ \sum_{s=1 \text{ to } k} \sum_{\text{all } t \neq s} \sum_{\text{all } u \neq s,t} \Pr\{\underline{E}_s \cap \underline{E}_t \cap \underline{E}_u\} + \ldots + (-1)^{k-1} \Pr\{\cap_{s=1-k} \underline{E}_s\}$$

Therefore the system reliability is

$$R = 1 - F$$

This method is applicable to any type of system if paths and cut sets are available, although disadvantageous in that all the relevant paths and cut sets must be known.

8.5.4   The Fratta–Montanari method

This method converts logical summation into algebraic summation (Fratta and Montanari, 1973). For example,

$$R = \Pr\{E_1 \cup E_2 \cup E_3 \cup \ldots\}$$

cannot directly produce an algebraic summation, so this is transformed, as shown below, so that the first term and the second term onward become exclusive events to each other:

$$R = \Pr\{E_1 + [\text{ not } E_1 \cap (E_2 \cup E_3 \cup \ldots)]\}$$

This transformation is repeated until the second term in square brackets becomes an empty event. This method also needs all the relevant paths and cut sets to be specified.

8.5.5   Path-and-cut method

If all the relevant paths and cut sets are available, they can be applied directly to the system structure, thereby reducing the amount of calculation more than does the enumeration method. If overlap occurs, Boolean algebra is required, which makes calculation complicated.

Of the methods described above for finding the system reliability, the path-and-cut method, which does not need the equivalent transformation of the system structure, is superior to the others in terms of simplicity of calculation. Nonetheless, for large or complicated systems, a tremendous amount of calculation is still needed, making it practically impossible to find the exact connection reliability. Hence heuristics are attractive, particularly where there is a trade-off between the effort expended and the accuracy achieved.

8.6   HEURISTIC METHODS

8.6.1   Method using all paths and cut sets

The exact value of system reliability can be obtained from

$$R' = E\left\{1 - \prod_{s=1\,\text{to}\,p}\left(1 - \prod_{i\in P(s)} x_i\right)\right\}$$

when all the paths are known (when $p = \boldsymbol{p}$), or from

$$R'' = E\left\{\prod_{s=1\,\text{to}\,k}\left(1 - \prod_{i\in K(s)}(1 - x_i)\right)\right\}$$

when all the cut sets are known (when $k = \boldsymbol{k}$). The calculation, however, needs Boolean algebra. R' is a monotonically increasing function in the number of paths p, and R'' is a monotonically decreasing function in the number of cut sets k. Curves A and B in Fig. 8.6 correspond to equations for R' and R'' respectively. When all paths are used, the true value of reliability is obtained at point a, whereas when all cut sets are used, the true value is obtained at point b.

Substituting x_i by r_i in both equations above, with the Boolean algebra omitted, produces the following upper and lower bounds:

$$U = 1 - \prod_{s=1\,\text{to}\,p}\left(1 - \prod_{i\in P(s)} r_i\right)$$

$$L = \prod_{s=1\,\text{to}\,k}\left(1 - \prod_{i\in K(s)}(1 - r_i)\right)$$

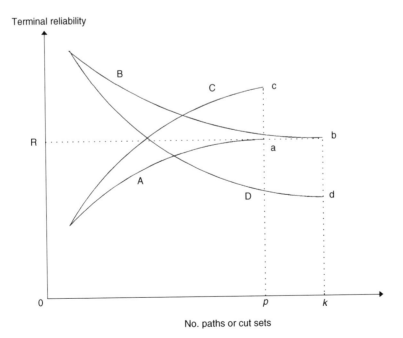

Figure 8.6 *Esary–Proschan upper and lower bounds*

When p is $\boldsymbol{p}$ and k is $\boldsymbol{k}$, U and L are known respectively as the Esary–Proschan upper and lower bounds for system reliability (Barlow and Proschan, 1965). The true value of system reliability lies between these bounds. In Fig. 8.6, curve C represents U for increasing numbers of paths p, with point c as the Esary–Proschan upper bound, and curve D represents L for increasing numbers of cut sets k, with point d as the Esary–Proschan lower bound.

8.6.2   Method using not all paths and cut sets

Table 8.1 shows what system reliability value the combination of the presence or absence of Boolean algebra and the number of paths and cut sets gives. Calculation is simplest and most efficient when Boolean algebra is omitted and when not all paths and cut sets are used. This method, however, cannot ensure that the value obtained is greater or smaller than the true value of system reliability or determine with what precision the reliability is estimated.

8.6.3   Intersection method

The curves C and D in Fig. 8.6 correspond to the case where Boolean algebra is omitted and where not all paths or cut sets are available. The intersection between these curves lies between the Esary–Proschan upper and lower bounds. The hypothesis is that the point of intersection constitutes a good approximation to the system reliability. If the intersection can be determined from a limited number of paths and cut sets, calculation is possible with a small amount of processing. Moreover, if the intersection offers a sufficiently good approximation to the true system reliability value, the heuristic is very efficient (Iida and Wakabayashi, 1990).

The intersection can be obtained at an early stage by choosing the sequence of paths and cut sets so that curve C and curve D are as steep as possible. First, find the path so that U is as great as possible. That is, determine the path s with the maximum value of $\prod_{i \in P(s)} r_i$. Note that

$$\ln \left(\prod_{i \in P(s)} r_i \right) = \sum_{i \in P(s)} \ln(r_i)$$

In this case, $0 \leqslant r_i \leqslant 1$, so $\ln(r_i) < 0$. When $-\ln(r_i) > 0$ is taken to be the length of link i, then finding the path that maximises U is equivalent to finding the shortest path.

Table 8.1   *Determination of system reliability*

	With Boolean algebra	Without Boolean algebra
All paths	$E\left\{1 - \prod_{s=1\,\text{to}\,p}\left(1 - \prod_{i\in P(s)} x_i\right)\right\}$	$1 - \prod_{s=1\,\text{to}\,p}\left(1 - \prod_{i\in P(s)} r_i\right)$
Partial paths	$E\left\{1 - \prod_{s=1\,\text{to}\,p}\left(1 - \prod_{i\in P(s)} x_i\right)\right\}$	$1 - \prod_{s=1\,\text{to}\,p}\left(1 - \prod_{i\in P(s)} r_i\right)$
All cut sets	$E\left\{\prod_{s=1\,\text{to}\,k}\left(1 - \prod_{i\in K(s)}(1 - x_i)\right)\right\}$	$\prod_{s=1\,\text{to}\,k}\left(1 - \prod_{i\in K(s)}(1 - r_i)\right)$
Partial cut sets	$E\left\{\prod_{s=1\,\text{to}\,k}\left(1 - \prod_{i\in K(s)}(1 - x_i)\right)\right\}$	$\prod_{s=1\,\text{to}\,k}\left(1 - \prod_{i\in K(s)}(1 - r_i)\right)$

This implies that to maximize the increase in curve C, all we have to do is find the n shortest paths.

To maximize the decrease in curve D, a *dual graph* is used. A dual graph is obtained by defining new nodes in regions surrounded by links and connecting all the adjacent new nodes. The length associated with each dual link is equal to the length of the primary link intersected (every dual link crosses one and only one primary link). The dual graph for the transportation network shown in Fig. 8.3 is given in Fig. 8.7. The cut sets between nodes i and j in Fig. 8.1 are represented by the paths between nodes i' and j' in the dual graph. Minimising L is equivalent to finding cut set s that maximises $\prod_{i \in K(s)} (1 - r_i)$. Note that

$$\ln \left(\prod_{i \in K(s)} (1 - r_i) \right) = \sum_{i \in K(s)} \ln(1 - r_i)$$

Again, $0 \leqslant 1 - r_i \leqslant 1$, so $\ln(1 - r_i) < 0$. When $-\ln(1 - r_i) > 0$ is taken to be the length of link i *in the dual graph*, then finding the path that minimises L is equivalent to finding the shortest path *in the dual graph*. This implies that to maximize the decrease in curve D, all we have to do is find the n shortest paths in the dual graph.

In the transportation network shown in Fig. 8.8, the links have uniform reliability values of either 0.9 (case 1) or 0.5 (case 2). Figure 8.9 gives the results of the intersection method for the connectivity reliability between nodes 1 and 16. In both cases, highly accurate values are obtained (an estimate of 0.1977 17 for a real reliability of 0.197505 in case 1, and an estimate of 0.197 32 for a real reliability of 0.198 44 in case 2). The intersection

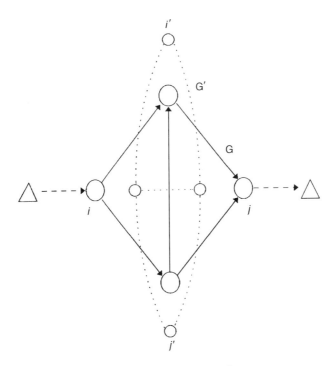

Figure 8.7 *Dual graph*

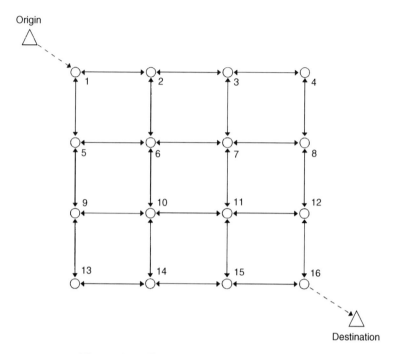

Figure 8.8 *Example transportation network*

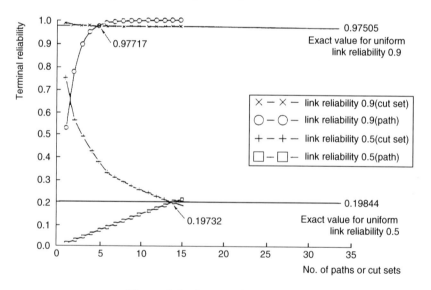

Figure 8.9 *Intersection method*

method is a heuristic. Nevertheless, it has been observed from a lot of numerical examples that this method can provide a good approximation of transportation network reliability.

One of the advantages of the intersection method is that an approximate value of connectivity reliability can be obtained efficiently. Another advantage is that the paths and cut sets used for the reliability calculation correspond to the more practical paths and cut sets from the perspective of transportation planning. That is, zigzag paths and distant detours, which would anyway be of little practical use, are considered last, and cut sets corresponding to cross sections or screen lines in the transportation network are chosen first.

For the approximation of system reliability, many methods other than the intersection method are also available. For instance, the inclusion-exclusion formula can be stopped midway when a given convergence standard is satisfied (Inoue, 1976). The larger the transportation network, however, the greater the amount of calculation needed, limiting the practical use of the method. A frequently used practical method is Monte Carlo simulation (see, for example, Kumamoto et al., 1977). Even with this method, many special techniques are needed to obtain accurate values and the amount of calculation is relatively great.

Recently, Du and Nicholson (1993) proposed a new approximation method for analysing transportation network reliability, taking into account the flow characteristics in addition to the topological structure of the network. This method makes full use of information about a number of most probable component state vectors (link capacity vectors), instead of solving the integrated equilibrium model of a degradable transportation system for all possible component state vectors. Using this algorithm, the connectivity reliabilities of a degradable transportation system can be determined with reasonable computation effort.

8.7 TRAVEL TIME RELIABILITY

Travel time reliability is the probability that a trip can reach its destination within a given period at a given time of day. In the case of commuting trips, the probability that a commuter can reach his work place from his home within a hour during the rush hour might be 80%. Thus the travel time reliability is a measure of the stability of travel time, and therefore is subject to fluctuations in flow. When the flow on a path does not fluctuate greatly, the destination can be reached more-or-less as scheduled. However, when flow fluctuation is large, travel time is often longer than expected. As levels of congestion in transportation networks grow generally, the stability of travel time will have greater significance to transportation users.

Assume that a path between a given pair of nodes (or an origin and a destination) consists of two links, 1 and 2. Past observations of flows on the links allow the distribution of travel time on links 1 and 2 to be established. The normal distribution is generally found to offer a reasonable description of travel time distributions. If the means and variances for links 1 and 2 respectively are μ_1, μ_2, σ_1^2 and σ_2^2, and if the distributions are normal, then the path travel time distribution is also normal with a mean given by the sum of the link means and a variance given by the sum of the link variances, namely

$$N(\mu_1 + \mu_2, \sigma_1^2 + \sigma_2^2)$$

provided the distributions are statistically independent. More generally, let T be the travel time on path s. Then

$$T \sim N\left(\sum_{i \in P(s)} \mu_i, \sum_{i \in P(s)} \sigma_i^2\right)$$

Noting that the variable

$$x = \left(T - \sum_{i \in P(s)} \mu_i \right) \bigg/ \sqrt{\sum_{i \in P(s)} \sigma_i^2}$$

has a unit normal distribution, the probability that travel time on path s is less than some threshold t may be found using the integral of the unit normal distribution, as follows:

$$\Pr\{T \leqslant t\} = \Phi\left(\left(t - \sum_{i \in P(s)} \mu_i \right) \bigg/ \sqrt{\sum_{i \in P(s)} \sigma_i^2} \right)$$

where $\Phi(y)$ is the area under the unit normal distribution from $-\infty$ to y.

Hence, unlike connectivity reliability, travel time reliability may be determined for individual paths (Asakura *et al.*, 1989).

8.8 FUTURE CHALLENGES

Analytical methods for transportation network reliability are needed to help construct and control transportation networks so that, even if some links in the network are congested or have failed, smooth travel can be ensured in the degraded network. Furthermore, network users need travel time stability.

The basic concepts and methods of transportation network reliability analysis have been described. Many practical problems remain to be solved. First, mutual relationships or dependencies among links are generally overlooked in studies of connectivity reliability. From the viewpoint of a network, however, when a link congests or fails, the adjacent links are also threatened by congestion or failure. Thus, in system reliability analysis, it is desirable to take the correlation among link reliabilities into account. Second, the flows on paths and cut sets are generally ignored when finding the connectivity reliability. Since paths and cut sets with heavy flows are highly important from the standpoint of traffic management, flow should be considered in reliability analysis in future studies. Third, travel time reliability differs from connectivity reliability in concept, and past studies are scarce and lack theoretical profundity. Further research is required here. Fourth, analytical methods for transportation network reliability in abnormal situations need to be developed. In situations such as disasters, accidents, and construction work, the flow variation differs greatly from that which prevails normally. Therefore it is impossible to apply the methods described so far to abnormal situations. If a part of a transportation network cannot be used due to an abnormal event, all the traffic previously using the affected link(s) should take other paths, with all the consequences this has for the reliability of other links. However, as one might expect, insufficient is known about flow variation in abnormal situations.

9

Network Design

9.1 INTRODUCTION

Network design is a very broad topic, spanning much of the material presented in earlier chapters. The topic concerns the configuration of transportation networks to achieve specified objectives. There are two forms of the problem, the *continuous network design problem* and the *discrete network design problem.*

The continuous network design problem takes the network topology as given and is concerned with the *parametrisation* of the network. Link parameters frequently considered include link (or node) *capacity* and *user charge.* Hence the following are examples of continuous network design problems:

1. the determination of road width (number of lanes);
2. the calculation of traffic signal timings; and
3. the setting of user charges (public transport fares, road tolls, etc.).

In general, there is a trade-off between user benefits on the one hand and the costs of the network alteration on the other. The construction of an extra lane increases the capacity of the road, leading to user benefits. However, the construction costs must be set against these user benefits. The capacity of a link can also be regulated by traffic signals. However, the allocation of more capacity to one signal-controlled link, through increased green time per cycle, must be matched by a reduction in the capacity of some other link(s), through reduced green time per cycle. The imposition of user charges can bring about benefits to society as a whole through greater economic efficiency, but may in certain circumstances result in extra costs for certain users.

The discrete network design problem is concerned with the *topology* of the network. The following are examples of discrete network design problems:

1. a road closure scheme;
2. the provision of a new public transport service (represented as a new set of links); and
3. the construction of a new road or rail link, perhaps a bridge, a tunnel or a bypass.

Here also there is a trade-off between user benefits arising from the addition or removal of a link and the costs of the network alteration.

The distinction between the continuous and discrete forms of the network design problem is not absolute, since the link capacity parameter of the continuous network design problem may be used to effectively *remove* links from the network. Furthermore, by considering *levels* of design parameter values, the continuous problem may be formulated as a discrete optimisation problem. It is generally thought that continuous problems are easier to solve than discrete ones, however the advent of probabilistic optimisation methods, in particular simulated annealing and genetic algorithms, may modify this view.

Traditionally network design has been concerned with the minimisation of *system cost* (equal to the sum of link flows times link costs) plus any additional costs incurred by the network alteration itself. However, this not only leads to ill-conditioned problems, in the sense that there are typically many design options leading to approximately the same total costs, but also leaves the benefits out of account. Where demand is elastic, *user surplus* would be a preferable objective for maximisation (user surplus is discussed in Chapter 1).

Network design involves the interaction of supply and demand. Oppenheim (1994) therefore describes the design process as a *bi-level programming problem*, the upper level problem being the *supply problem* and the lower level problem being the *demand problem*. The system designer *leads*, taking into account how the users *follow*. If the leader has prior knowledge of the responses of the followers, the situation is known in game theory as a *Stackelberg game* (see Fisk, 1986).

Various approaches to solving the network design problem have been formulated. An early heuristic approach involves iterating between the higher and lower level problems, solving the supply problem while holding the variables of the demand problem fixed, and then solving the demand problem while holding the variables of the supply problem fixed, etc. This iterative design-assignment algorithm was originally proposed by Allsop (1974) and Gartner (1976), probably independently, for traffic control problems, and by Steenbrink (1974) for other kinds of network design problem.

Since the solution of one problem only takes into account the solution of the other *in the previous iteration*, the result of the interactive design-assignment algorithm is a *Cournot–Nash game*, rather than a Stackelberg game as preferred. Even when both the upper level and the lower level consist of convex programming problems, the network design problem may be non-convex in the sense that there may be many local optima. The solution to the two kinds of game are therefore not always the same (see Fisk, 1986). That the iterative design-assignment method does not always yield the global optimal network design has been illustrated by Tan *et al.* (1979) and demonstrated theoretically by Marcotte (1981). An interesting variation on the Cournot–Nash game, proposed by Suwansirikul *et al.* (1987), optimises the design parameters for each link in series, treating the other design parameters as fixed. After each optimisation, the demand problem is resolved.

A second approach is to seek optimal design parameters directly using a pattern search method, such as the Powell or Hooke–Jeeves method, as proposed by Abdulaal and LeBlanc (1979). This approach is computationally intensive, because frequent evaluations of the deterministic (or stochastic) user equilibrium traffic assignment are required. Additionally, because of the non-convexity of the network design problem, difficulties may be encountered.

A third approach, proposed by Tan *et al.* (1979), is to replace the demand problem by its optimality conditions in the form of non-linear inequalities, and then incorporate these into the supply problem as additional constraints. However, this generally produces a non-convex feasible region making the application of standard constrained optimisation techniques impractical.

A fourth approach is based on the sensitivity of the primal and dual variables of the demand problem to perturbations of the design parameters. Sensitivity expressions have been derived for general non-linear programming problems by Fiacco (1976) and for the deterministic user equilibrium assignment by Tobin and Friesz (1988). The sensitivity expressions, which constitute a linear approximation to the demand problem *at the current solution*, are then taken as additional constraints for the supply problem, again producing a one-level problem. A sequence of one-level problems can then be solved, recalculating the sensitivity expressions for the latest solution on each iteration. Convergence is not guaranteed. Kim (1989) has applied this approach to network design.

As already noted, the network design problem is generally a non-convex programming problem. The constraints imposed by the demand problem lead to a non-convex feasible region. A *probabilistic search method*, such as simulated annealing or a genetic algorithm, would therefore be appropriate when a global optimum is sought. A fifth approach, proposed by Friesz *et al.* (1992, 1993), therefore makes use of simulated annealing. In Friesz *et al.* (1992), simulated annealing is applied to solve a bi-level programming formulation of the continuous network design problem. This necessitates frequently solving the demand problem, so regenerating a path set each time (as would be done by the Frank–Wolfe algorithm) is inefficient. Friesz *et al.* (1992) therefore suggest the prior enumeration of all paths likely to be used and then the solution of the demand problem by the projection method proposed by Bertsekas and Gafni (1982). In Friesz *et al.* (1993), simulated annealing is applied to a one-level problem with the demand side incorporated into the supply problem by the Tan *et al.* (1979) user equilibrium constraints (relaxed slightly). Genetic algorithms have yet to be reported in the context of optimal transportation network design.

9.2   OBJECTIVE FUNCTION

The user benefit is given by the sum of the integrals of the inverse demand functions, namely

$$UB = \sum_{j \in \text{OD}} \int_{x=0}^{x=t_j(z_j)} t_j^{-1}(x)\,dx$$

where $t_j^{-1}(x)$ is the inverse demand function for trip table element j, which is equal to the corresponding cost of travel, and OD is the set of trip table elements. The upper limit of integration is $t_j(z_j)$, which is the demand for trip table element j expressed as a function of the corresponding expected minimum trip cost. Note that z_j is itself a function of s through the path costs and path flows. If there is no cost at which an origin–destination flow would be zero, the user benefit is undefined. If demand is inelastic, then $t_j(z_j)$ is equal to some fixed value t_j , the range of integration is zero, and UB is zero.

The system cost is obtained by summing the products of link costs and link flows

$$SC = \sum_{i \in I} v_i c_i(v_i, s_i)$$

where s_i is the value of the design parameter for link i and v_i is the flow on link i (separable link cost functions are presupposed here).

In addition to the system cost, the cost of altering any design variable, sometimes referred to as the *investment cost*, must also be allowed for. The investment cost IC depends on both the size of the changes to the design parameters s - s^0 and the starting point s^0. Investment cost should always be non-negative, as both increasing and reducing capacity imply investment of some kind. For the sake of illustration, we shall assume that

$$IC = \beta \sum_{i \in I} |s_i - s_i^0| \geqslant 0$$

where β is the unit construction (or destruction) cost.

The *network surplus* is obtained by subtracting the system cost and the investment cost from the user benefit as follows:

$$\text{NS} = \text{UB} - \text{SC} - \text{IC} = \sum_{j \in \text{OD}} \int_{x=0}^{x=t_j(z_j)} t_j^{-1}(x)\,dx - \sum_{i \in I}(v_i, c_i(v_i, s_i) + \beta|s_i - s_i^0|)$$

Note that network surplus has the following partial derivatives with respect to trip table demand, link flow and link capacity:

$$\partial \text{NS}/\partial \mathbf{t} = \mathbf{z}^{\text{T}}$$

$$\partial \text{NS}/\partial \mathbf{v} = -\mathbf{c}^{\text{T}} - \mathbf{v}^{\text{T}}\mathbf{J}$$

$$\partial \text{NS}/\partial s_i = \begin{cases} -v_i\,\partial c_i/\partial s_i - \beta & \text{if } s_i > s_i^0 \\ -v_i\,\partial c_i/\partial s_i + \beta & \text{if } s_i < s_i^0 \end{cases}$$

where $\mathbf{J}$ is the Jacobian of the link cost functions with respect to link flows. Normally this would be required to be positive definite and invertible.

9.3 BI-LEVEL PROGRAMMING

9.3.1 Iterative design assignment

If for the moment we hold the link flows $\mathbf{v}$ fixed, NS is maximised when

$$\nabla \text{NS}(\mathbf{s}) = \mathbf{0}^{\text{T}}$$

provided no restrictions are placed on the feasible values of $\mathbf{s}$. Given the solution $\mathbf{s}^*$, an (possibly elastic) assignment algorithm may be used to obtain new link flows $\mathbf{v}$. This forms the basis of the iterative design-assignment bi-level programming algorithm.

Iterative design-assignment algorithm

Upper level   Solve the network design problem for $\mathbf{s}^*$ given $\mathbf{v}$. Proceed to the lower level.

Lower level   Given $\mathbf{s}^*$ find new $\mathbf{v}$. Return to the upper level.

The effect of the design variables $\mathbf{s}$ on the equilibrium link flows $\mathbf{v}$ and expected minimum trip table costs $\mathbf{z}$ depends on the assignment mechanism (the lower level problem). Where origin–destination demand is inelastic, the objective function reduces to minus the system cost and the investment cost. This is the objective most frequently used in practice.

In the Cournot–Nash game, the designer reacts to the users and the users react to the designer *in sequence*. The result of this game is known in general not to be the result of a Stackelberg game, whereby the designer takes into account the user response in the design process, because of the non-convex nature of the network design problem. Unfortunately, it is the Stackelberg game that is of greater practical interest in this context.

9.3.2   Sensitivity-based algorithm

By the chain rule of differentiation

$$dNS/ds = (\partial NS/\partial t)(\partial t/\partial z)(dz/dc)(\partial c/\partial s)$$
$$+ (\partial NS/\partial v)(dv/ds) + (\partial NS/\partial s)$$

and by substitution

$$dNS/ds = z^T(\partial t/\partial z)(dz/dc)(\partial c/\partial s)$$
$$+ (-c^T - v^T J)(dv/ds) + (\partial NS/\partial s)$$

where the trip table **t** and the link costs **c** are intermediate variables. The matrix of sensitivities (**dv/ds**) introduces the effect of reassignment approximately by making a linear approximation of the assignment function. Provided **s** has no restrictions, the condition

$$dNS/ds = 0^T$$

characterises the optimum design (if it exists). This leads to a modified bi-level programming problem, this time making use of the sensitivity of the assignment **v** to the design variables **s**.

Iterative sensitivity algorithm

Upper level   Solve the network design problem for s^* given **v** and
$z^T(\partial t/\partial z)(dz/dc)(\partial c/\partial s) + (-c^T - v^T J)(dv/ds)$. Proceed to the
lower level.

Lower level   Given s^* find new **v** and $z^T(\partial t/\partial z)(dz/dc)(\partial c/\partial s)$
$+ (-c^T - v^T J)(dv/ds)$. Return to the upper level.

The sensitivity of the equilibrium link flows and expected origin–destination costs to the design parameters depends on the assignment mechanism. This aspect is discussed in more detail in the following sections for the case of inelastic demand.

9.4   SENSITIVITIES

9.4.1   Deterministic user equilibrium assignment

The Tobin and Friesz (1988) approach to the analysis of sensitivities extends to the consideration of variations in costs as well (see Chapter 5). At optimality, the following relationships hold for the effect of small changes of the reduced primal and dual variables, Δh_u and Δz respectively, on the path costs and the trip table

$$\begin{bmatrix} \Delta g_u \\ \Delta t \end{bmatrix} = \begin{bmatrix} \nabla^2 f(h^*)_u & B_u^T \\ B_u & 0 \end{bmatrix} \begin{bmatrix} \Delta h_u \\ \Delta z \end{bmatrix}$$

where subscript u indicates that the set of paths has been reduced to include only those that are *used*. This is the **first order** effect of perturbations. Note that

$$A_u J(v^*) A_u^T = \nabla^2 f(h^*)_u$$

where $\mathbf{J}(\mathbf{v}^*)$ is the Jacobian of the link cost functions evaluated at the equilibrium link flows $\mathbf{v}^*$. The matrix between the two perturbation vectors is referred to as the *Jacobian of the system*. The inverse of this Jacobian provides the required sensitivities.

Writing the inverse Jacobian in the corresponding block form

$$\begin{bmatrix} \Delta \mathbf{h}_u \\ \Delta \mathbf{z} \end{bmatrix} = \begin{bmatrix} \mathbf{J}_{11} & \mathbf{J}_{12} \\ \mathbf{J}_{21} & \mathbf{J}_{22} \end{bmatrix} \begin{bmatrix} \Delta \mathbf{g}_u \\ \Delta \mathbf{t} \end{bmatrix}$$

it can be shown (see Chapter 5) that

$$\mathbf{J}_{11} = (\nabla^2 f(\mathbf{h}^*)_u)^{-1}(\mathbf{I} - \mathbf{B}_u{}^T(\mathbf{B}_u(\nabla^2 f(\mathbf{h}^*)_u)^{-1}\mathbf{B}_u{}^T)^{-1}\mathbf{B}_u(\nabla^2 f(\mathbf{h}^*)_u)^{-1})$$

$$\mathbf{J}_{12} = (\nabla^2 f(\mathbf{h}^*)_u)^{-1}\mathbf{B}_u{}^T(\mathbf{B}_u(\nabla^2 f(\mathbf{h}^*)_u)^{-1}\mathbf{B}_u{}^T)^{-1}$$

$$\mathbf{J}_{21} = (\mathbf{B}_u(\nabla^2 f(\mathbf{h}^*)_u)^{-1}\mathbf{B}_u{}^T)^{-1}\mathbf{B}_u(\nabla^2 f(\mathbf{h}^*)_u)^{-1}$$

$$\mathbf{J}_{22} = -(\mathbf{B}_u(\nabla^2 f(\mathbf{h}^*)_u)^{-1}\mathbf{B}_u{}^T)^{-1}$$

Hence where demand is inelastic, the following sensitivities may be calculated:

$$d\mathbf{v}/d\mathbf{s} = \mathbf{A}_u \mathbf{J}_{11} \mathbf{A}_u^T(\partial \mathbf{c}/\partial \mathbf{s})$$

$$d\mathbf{z}/d\mathbf{c} = \mathbf{J}_{21} \mathbf{A}_u^T$$

9.4.2   Stochastic user equilibrium

The network design problem incorporating stochastic user equilibrium assignment was originally looked at by Chen and Alfa (1992) and Davis (1993). Davis (1993) applied the logit path choice equations as non-linear constraints and looked for an exact local solution to the continuous network design problem. However, a different approach is adopted here.

Following Lam *et al.* (1996), consider the vector valued function

$$\mathbf{f}(\mathbf{v}, \mathbf{s}) = \mathbf{v} - \mathbf{P}(\mathbf{c}(\mathbf{v}, \mathbf{s}))\mathbf{t}$$

where $\mathbf{z}$ is a function of $\mathbf{c}(\mathbf{v}, \mathbf{s})$. At equilibrium, this function has value zero, so

$$\mathbf{f}(\mathbf{v}^*, \mathbf{s}) = \mathbf{0}$$

It is possible to infer how equilibrium link flows change with the link design parameters as follows. Note that

$$d\mathbf{v}/d\mathbf{s} = (\partial \mathbf{f}(\mathbf{v}^*, \mathbf{s})/\partial \mathbf{v})^{-1}(\partial \mathbf{f}(\mathbf{v}^*, \mathbf{s})/\partial \mathbf{s})$$

provided of course that the matrix $(\partial \mathbf{f}(\mathbf{v}^*, \mathbf{s})/\partial \mathbf{v})$ is invertible. Regarding the right-hand side of the above equation,

$$\partial \mathbf{f}(\mathbf{v}^*, \mathbf{s})/\partial \mathbf{v} = \mathbf{I} - \sum_{\text{all } j} t_j(\partial \mathbf{p}_j/\partial \mathbf{c})\mathbf{J}$$

where $\mathbf{p}_j$ is the jth column of the matrix of link choice proportions $\mathbf{P}$.

In the case of the logit path choice model with deterrence parameter α it may be readily verified that

$$\partial p_{ij}/\partial c_{i'} = \begin{cases} -\alpha p_{ij} + \alpha p_{ij}^2 & \text{if } i = i' \\ -\alpha p_{ii'j} + \alpha p_{ij} p_{i'j} & \text{otherwise} \end{cases}$$

where $p_{ii'j}$ is the proportion of traffic from the jth trip table element using both links i and i'.

Concerning the second term on the right-hand side,

$$\partial\mathbf{f}(\mathbf{v}^*, \mathbf{s})/\partial\mathbf{s} = -(\sum_{\text{all } j} t_j(\partial\mathbf{p}_j/\partial\mathbf{c})(\partial\mathbf{c}/\partial\mathbf{s})$$

where $p_{ii'j}$ is the proportion of traffic from the jth trip table element using both links i and i'.

Concerning the second term on the right-hand side

$$\partial\mathbf{f}(\mathbf{v}^*, \mathbf{s})/\partial\mathbf{s} = -\left(\sum_{\text{all } j} t_j(\partial\mathbf{p}_j/\partial\mathbf{c}) - \mathbf{P}(\partial\mathbf{t}/\partial\mathbf{z})\mathbf{P}^{\mathrm{T}}\right)(\partial\mathbf{c}/\partial\mathbf{s})$$

Bell (1995b) offers a way to evaluate both p_{ij} and $p_{ii'j}$. Let

$$w_{mn} = \begin{cases} \exp(-\alpha c_{mn}) & \text{if a link connects node or centroid } m \text{ to node or centroid } n \\ 0 & \text{otherwise} \end{cases}$$

The probability of travelling between any node or centroid m and any node or centroid n by *any* path is proportional to the corresponding element (namely $.v_{mn}$) of the following matrix:

$$\mathbf{N} = \mathbf{W} + \mathbf{W}^2 + \mathbf{W}^3 + \ldots = (\mathbf{I} - \mathbf{W})^{-1} - \mathbf{I}$$

No paths are excluded, so for networks with cycles the number of paths in the path set will be infinite. Furthermore, all paths will be used. The convergence of the above geometric series requires that $\mathbf{W}^n$ tends to zero as n tends to infinity. This may place a lower limit on the dispersion parameter, depending on the network topology.

The link choice proportions may be calculated as follows:

$$p_{imn} = v_{mr} \exp(-\alpha c_i) v_{r'n}/v_{mn}$$

where link i connects node or centroid r to node or centroid r' and $v_{rr'} = 1$ when $r = r'$. In a similar way

$$p_{ii'mn} = v_{mr} \exp(-\alpha c_i) \exp(-\alpha c_{i'}) v_{r''n}/v_{mn}$$

where link i' connects node or centroid r' to node or centroid r''.

9.5 NUMERICAL EXAMPLE

9.5.1 The network

To illustrate the problem of network design, consider the network of Fig. 2.9. Demand is assumed to be inelastic. The link cost functions are

$$c_1 = 1 + 2(v_1/s_1)^2$$
$$c_2 = 10 + 1(v_2/s_2)^2$$
$$c_3 = 1 + 2(v_3/s_3)^2$$
$$c_4 = 10 + 1(v_4/s_4)^2$$
$$c_5 = 1 + 2(v_5/s_5)^2$$

This network has three paths. Link 4 is particular to path 1, link 2 is particular to path 2 and link 3 is particular to path 3. The link capacities s_1, s_2, s_3, s_4 and s_5 are the design parameters. Each link is assumed to have a capacity of 3.2 in flow units (say vehicles per second) initially. This is a notional capacity which can be exceeded.

Links 1, 3 and 5 are relatively short but sensitive to the degree of saturation (the flow-to-capacity ratio). Links 2 and 4 by contrast are relatively long and insensitive to the degree of saturation. This network exhibits Braess's paradox, namely the expected travel time from the origin to the destination is reduced when link 3 is omitted from the network. When the total demand is 10 units of flow, the expected cost from the origin to the destination with link 3 in place *under a deterministic user equilibrium assignment* is 20.45 cost units (the centroid connectors are assumed to have zero costs). When link 3 is removed the expected origin to destination cost falls to 19.25 cost units, a saving of 1.2 cost units. Given an inelastic demand of 10 trips, this implies a total saving of 12 cost units. Taking $\beta = 1$ in the investment cost function, the cost of dismantling link 3 would be 5.12 cost units, leaving a net saving of 6.88 cost units.

Tables 9.1 and 9.2 show the sensitivity matrices for deterministic user equilibrium assignment. Note that as the cost of path 3 increases the expected minimum trip table cost falls, which is indicative of Braess's paradox. Tables 9.3 and 9.4 show the corresponding sensitivities relating to stochastic user equilibrium assignment.

Table 9.1 *Sensitivity of DUE path flow to path cost*

Flow/Cost	Path 1	Path 2	Path 3
Path 1	−0.2692	0.0518	0.2174
Path 2	0.0518	−0.2692	0.2174
Path 3	0.2174	0.2174	−0.4348

Table 9.2 *Sensitivity of DUE expected trip table cost to path cost*

	Path 1	Path 2	Path 3
Expected O–D cost	0.6667	0.6667	−0.3333

Table 9.3 *Sensitivity of SUE path flow to path cost*

Flow/Cost	Path 1	Path 2	Path 3
Path 1	−0.2215	0.0654	0.1561
Path 2	0.0654	−0.2215	0.1561
Path 3	0.1561	0.1561	−0.3123

Table 9.4 *Sensitivity of SUE expected trip table cost to path cost*

	Path 1	Path 2	Path 3
Expected O–D cost	0.6192	0.6192	−0.2385

As with deterministic user equilibrium assignment, an increase in the cost of path 3 reduces the expected minimum trip cost. Thus Braess's paradox might also apply to stochastic user equilibrium assignment. However, in this example network, removal of link 3 (and thereby also path 3) increases the expected trip cost from 22.0198 to 23.3789 cost units.

9.5.2 Iterative assignment algorithm

In this example, demand is assumed to be inelastic. The gradient of the objective function with respect to the network design parameters s *without considering reassignment* is therefore

$$\nabla NS(s) = -v^T(\partial c/s) - \beta 1^T$$

In this case

$$\partial c_i/\partial s_i = -2 \, b_i \, v_i^2/s_i^3$$

where b_i is the second (non-intercept) parameter of the link cost function for link i. Thus

$$\partial NS/\partial s_i = 2 \, b_i \, v_i^3/s_i^3 - \beta$$

If

$$s_i < \exp(\ln(2 \, b_i \, v_i^3/\beta)/3)$$

then

$$\partial NS/\partial s_i > 0$$

so s_i should be increased. If

$$s_i > \exp(\ln(2 \, b_i \, v_i^3/\beta)/3) > s_i^0$$

then

$$\partial NS/\partial s_i < 0$$

so s_i should be reduced. Hence NS is maximised when

$$s_i = \text{Max}\{s_i^0, \exp(\ln(2 \, b_i \, v_i^3/\beta)/3)\}$$

Initially every link was assumed to have a capacity of 3.12 units of flow. The iterative design-assignment algorithm converged after four iterations. The optimal allocations of capacity for each of the four iterations are shown in Table 9.5.

Table 9.5 *DUE capacity allocations by the iterative assignment algorithm*

	Link 1	Link 2	Link 3	Link 4	Link 5
Iteration 1	9.4450	5.1027	3.200	5.1027	9.4450
Iteration 2	11.9849	3.200	8.0164	3.200	11.9055
Iteration 3	15.8740	3.200	15.8740	3.200	15.8740
Iteration 4	15.8740	3.200	15.8740	3.200	15.8740

Table 9.6   *DUE expected O–D cost and objective function value*

	Expected O–D cost	NS before redesign	NS after redesign
Iteration 1	20.4456	205.1974	140.7828
Iteration 2	13.4627	151.4209	130.9222
Iteration 3	11.9157	141.4634	121.8330
Iteration 4	8.3811	121.8330	121.8330

The expected trip cost on each iteration, as well as the value of the objective function before and after the optimisation of the design, are shown in Table 9.6.

Iterations continue until no further design improvements can be made and deterministic user equilibrium assignment prevails. Comparison of the value of the objective function value after reassignment (NS before design) and after redesign (NS after redesign) shows zig-zag behaviour. In the early iterations, the reassignment leads to a deterioration in the objective function. Note that the iterative design-assignment algorithm will build capacity where demand is highest, which is not independent of the initial capacities.

When stochastic rather than deterministic user equilibrium assignment is assumed (taking $\alpha = 0.1$), the results are broadly similar. The capacity allocations are shown in Table. 9.7. While under deterministic user equilibrium assignment, all flow becomes concentrated on path 3, under stochastic user equilibrium assignment, all paths continue to be used. The expected trip cost on each iteration, as well as the value of the objective function before and after the optimisation of the design, are shown in Table 9.8. The iterations lead to a point where no design improvements can be made and the stochastic user equilibrium assignment prevails.

An interesting variation on the iterative design-assignment algorithm, which it is claimed approximates the Stackelberg game, has been proposed by Suwansirikul *et al.* (1987). The design parameter for each link is optimised in series, treating the other design parameters as fixed. The optimisation requires the repeated solution of the demand problem. The heuristic takes into account something of the Stackelberg leader–follower structure by evaluating the reaction of the network users to the design changes after the

Table 9.7   *SUE capacity allocations by the iterative assignment algorithm*

	Link 1	Link 2	Link 3	Link 4	Link 5
Iteration 1	10.1853	4.5152	4.4065	4.5152	10.1853
Iteration 2	11.2696	3.6545	6.6652	3.6545	11.2696
Iteration 3	11.4924	3.4777	7.1108	3.4777	11.4924
Iteration 4	11.5260	3.4510	7.1781	3.4510	11.5260
Iteration 5	11.5310	3.4471	7.1879	3.4471	11.5310

Table 9.8   *SUE expected O–D cost and objective function value*

	Expected O–D cost	NS before redesign	NS after redesign
Iteration 1	22.0198	219.6179	137.8502
Iteration 2	11.9945	136.2179	133.5774
Iteration 3	11.5094	132.7835	132.6993
Iteration 4	11.4551	132.5686	132.5668
Iteration 5	11.4480	132.5474	132.5474

Table 9.9 *DUE expected O–D cost and objective function value*

	Expected O–D cost	NS before redesign	NS after redesign
Iteration 1	20.4456	205.1974	159.8394
Iteration 2	14.2150	159.8394	121.9107
Iteration 3	8.4762	121.9107	121.8330
Iteration 4	8.3811	121.8330	121.8330

Table 9.10 *SUE expected O–D cost and objective function value*

	Expected O–D cost	NS before redesign	NS after redesign
Iteration 1	22.0198	219.6179	135.0764
Iteration 2	11.9981	135.0762	132.6422
Iteration 3	11.4771	132.6422	132.5527
Iteration 4	11.4493	132.5527	132.5450
Iteration 5	11.4470	132.5450	132.5441
Iteration 6	11.4468	132.5441	132.5441

design of each link has been optimised rather than after the design of the network as a whole has been optimised.

When applied to the example network, this heuristic appears to converge better in the first few iterations than the iterative design-assignment algorithm explored in the preceding Section. Convergence for deterministic user equilibrium assignment is shown in Table 9.9. The improvement in convergence speed is more marginal in the case of stochastic user equilibrium assignment (Table 9.10). In both cases, the improved convergence per iteration is achieved at the expense of considerable extra computation. The end result in both the deterministic and stochastic user equilibrium cases, however, is the same as obtained by the iterative design-assignment method, which corresponds to the Cournot–Nash game.

9.5.3 Iterative sensitivity algorithm

The effect of a small change in the design parameter for link i, namely s_i, on NS taking the effect of reassignment approximately into account is given by

$$dNS/ds_i = (-\mathbf{c}^T - \mathbf{v}^T\mathbf{J})(d\mathbf{v}/dc_i)(dc_i/ds_i) + (\partial NS/\partial s_i)$$
$$= (-\mathbf{c}^T - \mathbf{v}^T\mathbf{J})(d\mathbf{v}/dc_i)(dc_i/ds_i) + 2\,b_i\,v_i{}^3/s_i{}^3 - \beta$$

where $d\mathbf{v}/dc_i$ is the vector of sensitivities of equilibrium link flows to the cost of link i. The new upper level problem therefore allows for reassignment by making a linear approximation to the assignment function. The lower level problem determines the equilibrium link flows and their sensitivities to link costs. For the example network used in the previous section, Table 9.11 compares the sensitivity-based algorithm with the iterative design-assignment algorithm. For these calculations, β was taken to be 3. The values presented are the converged values where both the upper and lower level problems are solved. Surprisingly, the sensitivity-based algorithm takes significantly more iterations to converge.

Table 9.11 *A comparison of two algorithms*

Capacity:	Iterative design-assignment algorithm		Iterative sensitivity algorithm	
	DUE	SUE	DUE	SUE
Link 1	11.0064	7.7915	7.1076	7.1987
Link 2	3.2000	3.2000	3.2000	3.2000
Link 3	11.0064	4.5765	3.2000	3.2110
Link 4	3.2000	3.2000	3.2000	3.2000
Link 5	11.0064	7.7915	7.1076	7.1987
NS	179.7867	161.8369	171.7639	163.2643

In the case of deterministic user equilibrium assignment, the sensitivity-based algorithm is able to produce a better design. This appears not to be the case for stochastic user equilibrium assignment. This result was checked by increasing the dispersion parameter α in the logit path choice model from 0.1 (the value used for Table 9.11) to 1.0 to approximate a deterministic user equilibrium assignment. The stochastic user equilibrium capacity allocation moves toward the deterministic user equilibrium values as does the objective function value.

These results serve to confirm the non-convexity of the network design problem. The iterative design-assignment and the iterative sensitivity-based algorithm have found different local optima with objective function values that are not very different but designs that are quite different. The difference in designs is most pronounced for deterministic user equilibrium assignment, where the iterative design-assignment algorithm has produced an extreme solution.

9.6 CONCLUSIONS

Network design may be regarded as one of the 'final frontiers' of transportation network analysis, bringing together many of the elements covered in the preceding chapters. The problem is difficult to solve because of its non-convex nature and the complexity of realistic transportation networks. On the other hand, the problem is universal, relating both to the planning of transportation infrastructure and to the control of traffic in networks. It is a central theme of traffic management.

The first approach to the solution of the network design problem described in detail, the iterative design-assignment iterative method, is relatively simple to implement and appears to converge rapidly, at least on the basis of the example presented here. An iterative sensitivity-based algorithm appears to take longer to converge, again on the basis of the example presented here. The two methods produce different local optima, differing slightly in terms of the objective function but significantly in terms of the design.

A more satisfactory approach to the network design problem requires the application of new probabilistic optimisation methods, notably simulated annealing or a genetic algorithm, in order to be able to select from among the apparently many local optima. This is the subject of ongoing research.

10
Conclusions

10.1 NETWORK EQUILIBRIUM

Any attempt to present analytical techniques for transportation networks is necessarily date stamped and slanted toward the interests and knowledge of the authors. The approach adopted in this book is an equilibrium one. As argued in Chapter 1, many issues are simplified when equilibrium is assumed to hold, since it is not necessary to consider behavioural mechanisms in detail. Of course, if transportation systems are not in equilibrium, or approximately so, the value of the equilibrium approach is diminished.

This book deals with two forms of equilibrium, namely deterministic user equilibrium, where all users are perfectly informed and make rational decisions, and stochastic user equilibrium, where users still make rational decisions but imperfectly perceive the actual costs. In reality, an assumption of omniscience is not only unrealistic but also often inappropriate, for example where the provision of traveller information is the subject of the analysis.

In addition to a gain in realism, the assumption of stochastic user equilibrium resolves a number of problems that bedevil deterministic user equilibrium. Under deterministic user equilibrium, all used paths connecting a particular origin to a particular destination will have equal cost, but the distribution of flows across those paths may not be determined. Under stochastic user equilibrium assignment, however, paths with equal costs will have equal probabilities of being selected. Flow should therefore be shared equally across paths with equal costs. Since expected path flows are uniquely determined, path flow estimation (a major subject of the book) becomes possible.

There is of course a feedback from the paths chosen to the costs on which the choices are based. In the deterministic user equilibrium case, this feedback is accommodated through the link cost functions, examined in Chapter 4. In the stochastic user equilibrium case, the feedback is treated deterministically in that the flows and costs are treated as fixed values. However, since costs are *randomly* misperceived, the path choices and the resulting path flows are stochastic variables. Thus the costs which are randomly misperceived are themselves stochastic variables, adding to the random variation in path flows.

The effect of the feedback from flows to costs rather depends, amongst other things, on the lags involved (are, for example, today's decisions based on yesterday's costs?). Simulation results relating to the logit path choice model, reported in Hazelton *et al.* (1996), suggest that, when a lagged feedback is introduced, a uni-modal probability distribution for the frequency of choice of any path becomes a bi-modal distribution, due to flip-flop behaviour. This serves to show that while equilibrium models may correctly estimate the macroscopic properties of flow, the microscopic detail (like flow frequency distributions) require a better understanding of decision-making mechanisms. This is a fertile area of current research.

The logit path choice model, which arises from rational decision-making behaviour when perceived path costs are identically and independently Gumbell distributed, is given prominence in this book. In reality, there is no evidence that perceived path costs have this rather esoteric statistical distribution. Morever, the assumption of independence is debatable where paths substantially overlap. If the source of the random misperception of cost occurs at the level of the link, then the perceived costs of paths with shared links will be positively correlated. As a result of this observation, Daganzo and Sheffi (1977) and Maher and Hughes (1995) have argued in favour of the probit path choice model where these correlations can be taken into account. However, if the source of variation is instead the value of time, then the variation is across individuals rather than across links. All path costs will then be correlated to a greater or lesser extent, not just paths with shared links. Leurent (1996) has examined this case, putting forward a probit path choice model.

While the logit path choice model may exhibit some bias toward paths with shared links, they are used extensively in this book because of their tractability. The tendency to overload shared links is to some extent counteracted by the effect of monotonically increasing link cost functions.

10.2 TRAFFIC ASSIGNMENT

The way in which network users choose paths has been an underlying theme of the book. The subject has been treated generally, potentially including mode choice as well as route choice. Of course, the sphere of decision-making is broader still. Trip-makers may decide to choose alternative destinations, to retime their trips or indeed to cancel their trips altogether. These broader effects have been accommodated approximately in the book through an allowance for elastic origin–destination demands. In the case of destination choice, reference has been made in Chapter 5 to combined distribution and assignment models. Work on these models, dating back to Evans (1976), has led to attractively simple mathematical programming formulations and solution algorithms. Recent work on dynamic assignment, not touched on in this book, has brought the subject of trip scheduling to the forefront.

In congested networks, time dependencies must be allowed for. In reality, the flow into a link can temporarily exceed the flow out of it, provided the excess can be stored as a queue. Steady state queue and delay formulae suggest that as the flow-to-capacity ratio (the traffic intensity) approaches one, the queue and delay tend to infinity. A more accurate portrayal of congested conditions therefore requires time-dependent formulae. This book presents a compromise approach based on steady state time slices. The queue at the end one time slice is carried forward to the next and is processed before the new arrivals. The capacity available to process new arrivals in each time slice is thereby reduced by the amount of the queues carried forward. When the new arrivals exceed the available capacity, a new queue is formed. The number of new arrivals on each link during the time slice is sensitive to the delay (and therefore to the queue).

This compromise approach allows steady state mathematical programming methods to be applied to a time-dependent situation. Since queues carried forward are processed *before* the new arrivals, and because new queues do not form unless the new arrivals exceed the available capacity, FIFO discipline is enforced at the junctions at the macroscopic level. However, the assumption of steady state conditions within the time slice implies 'market clearing', namely that all trips are completed within the time slice, while the queues carried over from one time slice to the next correspond to incomplete trips. Nonetheless, provided the time slices are not too small, the approach probably provides a reasonably accurate description of the evolution of congestion over a peak period.

In both design and control applications, it is useful to know the sensitivity of the fitted values to the inputs, in this case the path costs and the trip table elements. Following the Tobin and Friesz (1988) approach, sensitivity expressions are developed to relate small changes in the inputs, in this case the trip table elements and path costs, to small changes in the primal and dual variables, in this case the path flows and the minimum origin-to-destination costs respectively. Expressions are presented both for the deterministic and for the stochastic user equilibrium assignment cases.

10.3 PATH FLOW ESTIMATION

A well-worn topic is the estimation of trip tables from traffic counts. The bulk of the research, and indeed most of the methods used in practice, assume proportional assignment (whereby a doubling of all the elements of a trip table would lead to a doubling of all link flows). This might not be a limitation for some kinds of networks, for example networks without route choice, or in situations where all links are counted. However, for the more usual case of a congested network where not all the links have been counted, proportional assignment is not appropriate. The introduction of non-proportional assignment leads to a bi-level programming problem, studied by Yang *et al.* (1992) and Yang (1995).

There are many ways to link the two levels of a bi-level programming problem. In Yang *et al.* (1992), the trip table is passed down from the upper (trip table estimation) level to the lower (traffic assignment) level and the assignment proportions are passed back. However, as noted earlier, the path flows and therefore the link choice proportions are in general not uniquely specified under deterministic user equilibrium assignment, causing stability problems. Moreover, this approach corresponds to solving a Cournot–Nash game, where Player A is minimising trip table variance (for example) while Player B is minimising the Beckman function to ensure a deterministic user equilibrium assignment. Neither player anticipates the response of the other. There is a risk of non-convergence. The game would be over more rapidly if Player A could anticipate (at least approximately) the actions of Player B. In Yang (1995), the sensitivities of the deterministic user equilibrium link flows to changes in the trip table are passed back, so the upper level problem can allow approximately for user responses. As one would expect, this approach is found to converge more rapidly.

As noted earlier, stochastic user equilibrium offers a more realistic description of macroscopic path choice behaviour. However, since path flows are uniquely identifiable under stochastic user equilibrium, these can be estimated directly from the traffic counts. In many applications, whether for control or for planning, it is in any case the path flows that are required. A path flow estimator is developed in Chapter 7, based on the logit path choice model. Sensitivity expressions relating small changes in the inputs, in this case the traffic counts and the path costs, to small changes in the primal and dual variables, in this case the path flows and the equilibrium delays respectively, are derived.

10.4 NETWORK RELIABILITY

The topic of transportation network reliability is relatively under-researched. Not surprisingly, the topic has received most attention in those countries most threatened with natural catastrophes, like Japan, New Zealand and Norway. It is, however, a topic of much wider interest and deserving much more attention. Ideally both individual and communal transport networks should be designed with a degree of redundancy in mind. The vulnerability of networks with respect to the possibility of failure of individual links may be quantified. Chapter 8 presents a number of approaches to the problem, but further research is justified.

An important element of network reliability is user behaviour. The capacity of a network, perhaps defined as the largest multiplier that may be applied to a trip table, will depend on the traffic assignment. The effect on capacity of a link failure will therefore also depend on the traffic assignment.

10.5 NETWORK DESIGN

This is a topic that potentially spans the whole field of transportation network analysis. It relates to the planning and design of transportation networks as well as to the control of traffic. Traffic signal control is in essence a network design problem. As user response is an important element of network design, a bi-level programming problem is generated. The upper level problem is concerned with network design so as to maximise some objective (such as network surplus) while the lower level problem is concerned with how users will react to network changes.

A range of solution approaches are to be found in the literature. The fundamental non-convexity of the problem, however, presents difficulties for mathematical programming methods. The current consensus appears to be that a probabilistic optimisation method, such as simulated annealing, is required to find a global solution. A combination is required of a bi-level programming method to find a local solution and a probabilistic search method to compare the local solutions. The application of probabilistic optimisation methods to the network design problem is in its infancy.

References

Abdulaal, M. and LeBlanc, L. (1979) Continuous equilibrium network design models. *Transportation Research*, **13B**, 19–32.

Addison, J. D. and Heydecker, B. G. (1993) A mathematical model for dynamic traffic assignment. *Proceedings of the 12th International Symposium on Transportation and Traffic Theory* (Ed. C. Daganzo), Elsevier: Amsterdam, pp 171–186.

Akamatsu,. (1996) Cyclic flows, Markov processes and stochastic traffic assignment. *Transportation Research B*, in press.

Allsop, R. E. (1974) Some possibilities for using traffic control to influence trip distribution and route choice. *Proceedings of the 6th International Symposium on Transportation and Traffic Theory* (Ed. D.J. Buckley), Elsevier: New York.

Allsop, R. E. (1992) Evolving application of mathematical optimisation in design and operation of individual signal-controlled road junctions. In: *Mathematics in Transport Planning and Control* (Ed. J. D. Griffiths), Clarendon Press: Oxford.

Asakura, Y., Kashiwadani, M. and Kumamoto, N. (1989): Reliability measures of regional road network due to daily variance in traffic flows, *Infrastructure Planning Review*, **7**, 235–242 (in Japanese).

Ashok, K. and Ben-Akiva, M. E. (1993) Dynamic origin–destination matrix estimation and prediction for real-time traffic management systems. In: *Transportation and Traffic Theory* (Ed. C.F. Daganzo), Elsevier: Amsterdam, pp 465–484.

Barlow, R. E. and Proschan, F. (1965) *Mathematical Theory of Reliability*, John Wiley & Sons: New York

Beckman, M. J., McGuire, C. B. and Winston, C. B. (1956) *Studies in the Economics of Transportation* Cowles Commission Monograph, Yale University Press: New Haven, CN.

Bell, M. G. H. (1985) Variances and covariances for origin–destination flows when estimated by log-linear models. *Transportation Research B*, **19B**, 497–507.

Bell, M. G. H. (1991) The estimation of origin–destination matrices by constrained generalised least squares. *Transportation Research B*, **25B**, 115–125.

Bell, M. G. H. (1995a) Stochastic user equilibrium assignment in networks with queues. *Transportation Research B*, **29B**, 125–137.

Bell, M. G. H. (1995b) Alternatives to Dial's logit assignment algorithm. *Transportation Research B*, **29B**, 287–295.

Bell, M. G. H. (1995c) A log-linear model for path flow estimation. In: *Applications of Advanced Technologies in Transportation Engineering* (Ed. Y. J. Stephanades and F. Filippi), American Society of Civil Engineering: New York, pp 695–699.

Bell, M. G. H., Lam, W. H. K., Ploss, G. and Inaudi, D. (1993) Stochastic user equilibrium assignment and iterative balancing. In: *Transportation and Traffic Theory* (Ed. C. F. Daganzo), Elsevier: Amsterdam, pp 427–440.

Bell, M. G. H., Lam, W. H. K. and Iida, Y. (1996) A time-dependent multiclass path flow estimator. *Proceedings of the 13th International Symposium on Transportation and Traffic Theory*, Lyon, July.

Ben-Akiva, M. and Lerman, S. (1985) *Discrete Choice Analysis: Theory and Application to Travel Demand*. MIT Press: Cambridge.

Bertsekas, D. P. and Gafni, E. M. (1982) Projection methods for variational inequalities with application to the traffic assignment problem. *Mathematical Programming Study*, **17**, 139–159.

Braess, D. (1968) Über ein Paradox in der Verkehrsplanung. *Unternehmensforschung*, **12**, 258–268.

Bruynooghe, M., Gibert, A. and Sakarovitch, M. (1968) Une mëthode d'affectation du trafic. *Proceedings of the 4th International Symposium on the Theory of Road Traffic Flow*, Karlsruhe.

Burrell, J. E. (1968) Multiple route assignment and its application to capacity restraint. *Proceedings of the 4th International Symposium on the Theory of Road Traffic Flow*, Karlsruhe.

Carey, M. (1992) Nonconvexity of the dynamic traffic assignment problem. *Transportation Research*, **26B**, 127–132.

Cascetta, E. (1984) Estimation of origin–destination matrices from traffic counts and survey data: A generalised least squares estimator. *Transportation Research*, **18B**, 289–299.

Cascetta, E., Inaudi, D. and Marquis, G. (1993) Dynamic estimation of origin–destination matrices using traffic counts. *Transportation Science*, **27**, 363–373.

Castillo, J. M. del, Pintado, P. and Benitez, F. G. (1994) The reaction time of drivers and the stability of traffic flow. *Transportation Research B*, **28B**, 35–60.

Chen, M. and Alfa, S. (1991) Algorithms for solving Fisk's traffic assignment model. *Transportation Research*, **25B**, 405–412.

Chen, M. and Alfa, S. (1992) A network design algorithm using a stochastic incremental traffic assignment approach. *Transportation Science*, **25**, 215–224.

Cremer, M. and Papageorgiou, M. (1981) Parameter identification for a traffic flow model. *Automatica*, **17**, 837–844.

Daganzo, C. F. (1982) Unconstrained extremal formulation of some transportation equilibrium problems. *Transportation Science*, **16**, 332–360.

Daganzo, C. F. (1995) Requiem for second-order fluid approximations of traffic flow. *Transportation Research B*, **29B**, 277–286.

Daganzo, C. F. and Sheffi, Y. (1977) On stochastic models of traffic assignment. *Transportation Science*, **11**, 253–274.

Davis, G. (1993) Exact local solution to the continuous network design problem via stochastic user equilibrium assignment. *Transportation Resarch B*, **27B**, 61–75.

Dial, R. B. (1971) A probabilistic multipath traffic assignment model which obviates the need for path enumeration. *Transportation Research*, **5**, 83–111.

Dijkstra, E. W. (1959) A note on two problems in connection with graphs. *Numerische Math.*, **1**, 269–271.

Downs, A. (1962) The law of peak-hour expressway congestion. *Traffic Quarterly*, **16**, 393–409.

Drissi-Kaitouni, O. (1993) A variational inequality formulation of the dynamic traffic assignment problem. *European Journal of Operational Research*, **71**, 188–204.

Drissi-Kaitouni, O. and Hameda-Benchekroun, A. (1992) A dynamic traffic assignment model and a solution algorithm. *Transportation Science*, **26**, 119–128.

Du, Zhen-Ping and Nicholson, A. J. (1993) Degradable transport systems: Performance, sensitivity and reliability analysis. Research Report 93-8, Department of Civil Engineering, University of Canterbury, Christchurch, New Zealand.

Evans, A. W. (1992) Road congestion pricing: When is it a good idea? *Journal of Transport Economics and Policy*, **26**, 213–243.

Evans, S. P. (1976) Derivation and analysis of some models for combining trip distribution and assignment. *Transportation Research*, **10**, 37–57.

Fellendorf, M. (1994) VISSIM: Ein Instrument zur Beurteilung verkehrsabhängiger Steuerung. In: Tagungsband zum Kolloquium 'Verkehrsabhängige Steuerung am Knotenpunkt', Forschungsgesellschaft für Strassen- und Verkehrswesen: Köln.

Fiacco, A. V. (1976) Sensitivity analysis for non-linear programming using penalty methods. *Mathematical Programming*, **10**, 287–311.

Fisk, C. (1980) Some developments in equilibrium traffic assignment. *Transportation Research*, **14B**, 243–255.

Fisk, C. (1986) A conceptual framework for optimal transportations system planning with integrated supply and demand models. *Transportation Science*, **20**, 37–47.

Floyd, R. W. (1962) Algorithm 97, shortest path. *Communications of the Association of Computing Machinery*, **5**, 345.

Ford, L. R. Jr and Fulkerson, D. R. (1962) *Flows in Networks*. Princeton University Press: Princeton, NJ.

Fratta, L. and Montanari, U. G. (1973) A Boolean algebra method for computing the terminal reliability in a communication network. *IEEE Trans. Circuit Theory*, CT-20, No. 3, 203–211.

Friesz, T. L., Cho, H.-J., Mehta, N. J., Tobin, R. L. and Anandalingam, G. (1992) A simulated annealing approach to the network design problem with variational inequality constraints. *Transportation Science*, **26**, 18–26.

Friesz, T. L., Anandalingam, G., Mehta, N. J., Nam, K., Shah, S. J. and Tobin, R. L. (1993) The multiobjective equilibrium network design problem revisited: A simulated annealing approach. *European Journal of Operational Research*, **65**, 44–57.

Gartner, N. H. (1976) Area traffic control and network equilibrium. In: *Lecture Notes in Economics and Mathematical Systems, Vol. 118* (Ed. M. A. Florian), Springer-Verlag: Berlin, pp 274–297.

Gartner, N. H. (1980a) Optimal traffic assignment with elastic demands: A review. Part I: Analysis framework. *Transportation Science*, **14**, 174–191.

Gartner, N. H. (1980b) Optimal traffic assignment with elastic demands: A review. Part II: Algorithmic approaches. *Transportation Science*, **14**, 192–208.

Hazelton, M., Lee, S. and Polak, J. (1996) Stationary states in stochastic process models of traffic assignment: a Markov Chain Monte Carlo method. *Proceedings of the Universities Transport Studies Group 28th Annual Conference*, Huddersfield, January (unpublished).

Henley, E. J. and Kumamoto, H. (1981) *Reliability Engineering and Risk Assesment*, Prentice Hall: Englewood Cliffs, NJ.

Heydecker, B. G. (1986) On the definition of traffic equilibrium. *Transportation Research*, **20B**, 435–440.

Hills, P. J. (1993) Road congestion pricing: When is it a good policy? A comment. *Journal of Transport Economics and Policy*, **27**, 91–99.

Hunt, P. B., Robertson, D. I., Bretherton, R. D. and Royle, M. C. (1982) The SCOOT on-line traffic signal optimisation technique. *Traffic Engineering and Control*, **23**, 190–192.

Iida, Y. and Wakabayashi H. (1990) An approximation method of terminal reliability of road network using patial minimal path and cut sets, *Proceedings of 5th WCTR*, Vol. 4, pp 367–380.

Inoue, K. (1976) System reliability and safety analysis. *Journal of Japan Society of Mechanical Engineering*, **79**, (686), 56–61 (in Japanese).

Kim, T. J. (1989) *Integrated Urban Systems Modeling: Theory and Applications*. Kluwer: Boston.

Kimber, R. M. and Daly, P. N. (1986) Time-dependent queueing at road junctions: Observation and prediction. *Transportation Research*, **20B**, 187–203.

Kimber, R. M. and Hollis, E. M. (1979) Traffic queues and delays at road junctions. TRRL Laboratory Report 909, Transport and Road Research Laboratory.

Köhl, W. (1994) Verkehrsvermeidung durch Stadt- und Landesplanung? *Straßenverkehrstechnik*, **38**, 251–260.

Kumamoto, H., Tanaka, K. and Inoue, K. (1977) Efficient evaluation of system reliability by Monte Carlo method, *IEEE Trans. Reliability*, **R-26**, 311–315.

Lam, W. H. K., Zhang, N. and Lo, H.P. (1995) Time varying stochastic traffic assignment model with simultaneous queues for multiple user classes. Submitted to *Transportation Research B*.

Lam, W. H. K., Zhang, N. and Lo, H. P. (1996) A transport network design tool with elastic demand for Pearl River Delta. Submitted to *Transportation Research B*.

Leonard, D. R., Tough, J. B. and Baguley, P. C. (1978) CONTRAM — A traffic assignment model for predicting flows and queues during peak periods. TRRL Laboratory Report LR841, Transport and Road Research Laboratory.

Leurent, F. M. (1996) The theory and practice of a dual criteria assignment model with a continuously distributed value of time. *Proceedings of the 13th International Symposium on Transportation and Traffic Theory*, Lyon, July.

Leutzbach, W. (1988) *Introduction to the Theory of Traffic Flow*. Springer-Verlag: Berlin.

Lighthill, M. J. and Whitham, G. B. (1955) On kinematic waves, II. A theory of traffic flow on roads. *Proceedings of the Royal Society*, **229A**, 317–345.

Lo, H. P., Zhang, N. and Lam, W. H. K. (1995) Decomposition algorithm for statistical estimation of OD matrix with random link choice proportions from traffic counts. *Transportation Research B*, **30B**, 309–324.

Maher, M. J. (1983) Inferences on trip matrices from observations on link volumes: a Bayesian statistical approach. *Transportation Research B*, **17B**, 435–447.

Maher, M. J. and Hughes, P. C. (1995) A probit-based stochastic user equilibrium assignment model. Submitted to *Transportation Research B*.

Marcotte, P. (1981) An analysis of heuristics for the continuous network design problem. *Proceedings of the 8th International Symposium on Transportation and Traffic Theory*. University of Toronto:Toronto, pp 452–468.

Meneguzzer, C. (1995) An equilibrium route choice model with explicit treatment of the effect of intersections. *Transportation Research B*, **29B**, 329–356.

Mine, H. and Kawai, H. (1982) *Mathematics for Reliability and Availability*, Asakura-shoten (in Japanese).

Mogridge, M. J. H. (1995) Modal equilibrium in congested urban networks. Submitted to *Transportation*.

Nakazawa, H. (1981): Decomposition method for computing the reliability of complex networks, *IEEE Trans action on Reliability.*, **R30**, 289–292.

Oppenheim, N. (1994) *Urban Travel Demand Modeling*. Wiley-Interscience: New York.

Ortuzar, J. de and Willumsen, L. G. (1990) *Modeling Transport*. Wiley: New York.

Pages, A. and Gondran M. (1986) *System Reliability — Evaluation and Prediction in Engineering.*, North Oxford Academic.

Papageorgiou, M., Blosseville, J.-M. and Hadj-Salem, H. (1989) Macroscopic modelling of traffic flow on the Boulevard Peripherique in Paris. *Transportation Research B*, **23B**, 29–47.

Patriksson, M. (1994) *The Traffic Assignment Problem: Models and Methods*. VSP: Utrecht.

Payne, H. J. (1979) FREEFLOW: A macrospic simulation model of freeway traffic. *Transportation Research Record*, **772**, 68–75.

Powell, W. and Sheffi, Y. (1982) The convergence of equilibrium algorithms with pre-determined step sizes. *Transportation Science*, **16**, 45–55.

RAC (1995) *Car Dependence*, RAC Foundation for Motoring and the Environment.

Samuelson, P. A. (1970) *Economics* (8th edn). McGraw-Hill Kogakusha: New York and Tokyo.

Sheffi, Y. (1985) *Urban Transportation Networks*. Prentice-Hall: Englewood Cliffs, NJ.

Sherali, H. D., Sivanandan, R. and Hobeika, A. G. (1994) A linear programing approach for synthesising origin–destination trip tables from link traffic volumes. *Transportation Research B*, **28B**, 213–234.

Shogan, A. W. (1978): A decomposition algorithm for network reliability analysis, *Networks*, **8**, 231–251.

Smith, M. J. (1979) The existence, uniqueness and stability of traffic equilibria. *Transportation Research*, **13B**, 293–304.

Smith, M.J. (1987) Traffic control and traffic assignment in a signal controlled road network with queueing. *Proceedings of the 10th International Symposium on Transportation and Traffic Theory* (Ed. H. Gartner and N.H.M. Wilson), MIT Press: Cambridge, pp 319–338.

Smith, T. E. (1987) A cost-efficiency theory of dispersed network equilibrium. *Environment and Planning*, **17A**, 688–695.

Steenbrink, P. A. (1974) *Optimisation of Transport networks*. Wiley: New York.

Suwansirikul, C., Friesz, T. L. and Tobin, R. L. (1987) Equilibrium decomposed optimisation: A heuristic for the continuous network design problem. *Transportation Science*, **21**, 254–264.

Tan, H. N., Gershwin, S. S. and Athans, M. (1979) Hybrid optimisation in urban traffic networks. Technical Report No. DOT-TSC-RSPA-79-7, National Technical Information Service Springfield, Ill.

Thompson, J. M. (1977) *Great Cities and their Traffic*. Gollancz: London.

Tobin, R. L. and Friesz, T. L. (1988) Sensitivity analysis for equilibrium network flow. *Transportation Science*, **22**, 242–250.

Van Vliet, D. (1978) Improved shortest path algorithms for transport networks. *Transportation Research*, **12**, 7–20.

Van Vliet, D. (1981) Selected node-pair analyis in Dial's assignment algorithm. *Transportation Research*, **15B**, 65–68.

Van Zuylen, H. J. and Willumsen, L. G. (1980) The most likely trip matrix estimated from traffic counts. *Transportation Research B*, **14B**, 281–293.

Wakabayashi, H., Iida, Y. and Inoue, Y.(1993) A link reliability estimation method using characteristics of traffic flow fluctuation in road network by computer simulation. *Journal of Infrastructure Planning and Management*, Japan Society of Civil Engineers, **4–29**, (58), 35–44 (in Japanese).

Wardrop, J. G. (1952) Some theoretical aspects of road traffic research. *Proceedingss of the Intitution of Civil Engineers*, Part II(1), pp 325–378.

Warshall, S. (1962) A theorem on Boolean matrices. *Journal of the Association of Computing Machinery*, **9**, 11–12.

Watling, D. (1996) Stability of a traffic assignment model. *Proceedings of the Universities Transport Studies Group 28th Annual Conference*, Huddersfield, January (unpublished).

Webster, F. V. (1958) Traffic signal settings. Road Research Laboratory Technical Paper 39.

Webster, F. V. and Cobbe, B. M. (1966) Traffic Signals. Ministry of Transport, Road Research Technical Paper No. 56, London, HMSO.

Wie, B. W., Friesz, T. L. and Tobin, R. L. (1990) Dynamic user optimal traffic assignment on congested multidestination networks. *Transportation Research B*, **24b**, 431–442.

Williams, H. C. W. L. (1977) On the formation of travel demand models and economic evaluation measures of user benefit. *Environment and Planning*, **9A**, 285–344.

Wilson, A. G. (1967) Entropy maximising models in the theory of trip distribution, mode split and route split. *Journal of Transportation Economic and Policy*, **3**, 108–126.

Yang, H. (1995) Heuristic algorithms for the bilevel origin-destination matrix estimation problem. *Transportation Research B*, **29B**, 231–242.

Yang, H. and Meng, Q. (1995) Departure time, route choice and congestion toll in a queueing network with elastic demand. Submitted to *Transportation Research*.

Yang, H. and Wong, S. C. (1995) Reserve capacity of a signal-controlled road network. Submitted to *Transportation Research*.

Yang, H., Yagar, S. and Iida, Y. (1994) Traffic assignment in a congested discrete/continuous transportation system. *Transportation Research B*, **28B**, 161–174.

Yang, H., Sasaki, T., Iida, Y. and Asakura, Y. (1992) Estimation of origin–destination matrices from link traffic counts on congested networks. *Transportation Research B*, **26B**, 417–434.

Zawack, D. J. and Thompson, G. L. (1987) A dynamic network space-time network flow model for city traffic congestion. *Transportation Science*, **21**, 153–162.

Zhang, J. and Smith, A. (1995) Trip assignment models with multiple user classes. Submitted to *Transportation Research B*.

Index